Pittsburgh Post-Gazette

PUBLISHER & EDITOR-IN-CHIEF: JOHN ROBINSON BLOCK
EDITOR AND VICE-PRESIDENT: JOHN G. CRAIG JR.
MANAGING EDITOR: MADELYN ROSS

Reporting and Writing: Caroline Abels, Bob Batz Jr., Michael A. Fuoco, Tom Gibb, John Hayes, L.A. Johnson, Cindi Lash, Steve Levin, Jim McKay, Johnna A. Pro, Milan Simonich, Anita Srikameswaran
Photography: John Beale, Franka Bruns, V.W.H. Campbell, Peter Diana, Lake Fong, Matt Freed, John Heller, Annie O'Neill, Martha Rial, Darrell Sapp, Bill Wade
Graphics: Ted Crow, James Hilston, Stacy Innerst, Daniel Marsula, Bill Pliske, Steve Thomas
Photo Editors: Dave LaBelle, Doug Oster
Editors: Madelyn Ross, Mark Roth

Special Thanks to the Quecreek Nine and their families, who graciously gave their time to the reporters, photographers and artists of the *Pittsburgh Post-Gazette.*

Triumph Books
601 South LaSalle Street
Suite 500
Chicago, Illinois, 60605
(312) 939-3330 Fax: (312) 663-3557

Printed in the United States of America

ISBN 1-57243-537-2

Front cover photograph by Marty Ginter
Back cover, inside front cover photographs by John Beale

Book design by

www.spectrumdg.com

Typefaces: Fenice, Humanist521, Century ITC, Impact

ALL NINE ALIVE!

The Dramatic Mine Rescue That Inspired and Cheered a Nation

FROM THE PAGES OF THE

Pittsburgh Post-Gazette

TRIUMPH BOOKS

CHICAGO

Contents

Introduction

Western Pennsylvania's Somerset County nestles in the beautiful Laurel Highlands, a swath of undulating farmlands, picturesque towns, and vacation cottages. Enveloped by fresh mountain air and carpeted with frilly ferns, its claim to fame had been its serenity, its only excitement borrowed from the city of Pittsburgh, 50 miles west.

Then came September 11, 2001, and a plane full of terrorists and heroes, all of whom disappeared in a dusty crater in an unlikely Somerset County field. The events of that day were so horribly indelible that Somerset became synonymous with stunning heroics and stunning loss—but mostly loss.

So on July 24, 2002, when word reached the newsroom that nine miners were trapped in a fast-flooding coal mine in Somerset County, we assumed the worst. Another national story and more loss, just 13 miles from the site of Flight 93. Indeed, for four days bad luck was trumped by worse luck as the rescue attempt sputtered and stalled, each agonizing hour filling a rapt world with more and more dread.

But you already know how this suspense ended. This time it was triumph, rescue, life, and talk of miracles. This time we were lifted up, crying for joy over nine men we didn't know, marveling at technology we didn't understand, cheering individual courage and selfless teamwork we rarely see, and celebrating a happy ending we dared not expect.

This book is the inspiring story of the miners and rescuers who brought us blessed relief from a national grief persisting since September 11. It's told in words, photographs, and graphics by nearly three dozen *Post-Gazette* journalists whose own teamwork and talents soared to match the moment.

—The Editors

CHAPTER ONE

THE MINERS

For thousands of years, sinuous stripes of bituminous coal have lain beneath the surface of the wooded hills and valleys of what is now Somerset County, Pennsylvania. Its extraction fueled an industrial revolution, lured our immigrant ancestors, and contributed to Western Pennsylvania's reputation for hard work and hard living.

As a piece of our history, coal mining has seemed herculean, monumental, even romantic.

But on the afternoon of Wednesday, July 24, it was just a job.

Eighteen miners left their homes in small towns dotting the Laurel Highlands and drove to Quecreek Mine, which lay beneath a dairy farm in Lincoln Township just off Somerset Pike. They gathered at its entry portal at 2:30 P.M., just as most of them had for five or six days a week since March.

There they split into two crews of nine, one to enter and head straight south, the other to bear left and begin chipping the east face.

With clouds rolling in and out, it was an agreeable day. But soon they would leave it behind, riding a motorized cart on a dug-out ramp a mile and a half long, which would take them into the cool darkness 245 feet below the surface—as far down as a 25-story building is up.

At 31, **Harry Blaine Mayhugh Jr.** was the youngest on his crew of nine. They rarely called him by his given names. He was "Stinky." He called them by equally affectionate nicknames.

Mayhugh was one of the guys. The living room of his home in Meyersdale, about 25 miles south of the Quecreek Mine, is decorated in a classic southwestern Pennsylvania sportsman's motif, including a mounted buck's head, fishing lures and other wildlife knickknacks, and ducks floating around the ceiling on the wallpaper border.

The husky six-footer played football and baseball at

Meyersdale High School. Just after he graduated in 1989, he started dating Leslie Foy, who was entering her junior year. They became engaged while he was in the U.S. Navy, and when his two years were up in 1992, they got married and had a son and a daughter.

Mayhugh worked in a factory and then for a lawn-care company before becoming a deep miner in 1997. Despite having to contort his big frame for eight hours in the 4- to-4½-foot-high mine shafts, he enjoyed the work—or more precisely, those he worked with. He relished the friendships, formed with men who were down-to-earth, family-oriented, and God-fearing. Separated from the world above, they had to rely on each other every day.

Leslie understood both the job's draw and its dangers. Her father, Thomas Foy, had been a coal miner since before she was born, and he and her husband now worked for the same outfit: Black Wolf Coal Company.

Since March 10, they'd been working on the same crew.

Every day before leaving for the mine, Mayhugh would give his wife a good-bye kiss.

But on this Wednesday she was still out back, finishing yard work they'd started together earlier, and he didn't have time. So he waved and called out, "I love you, honey." Then he was gone.

Thomas Foy, Mayhugh's 52-year-old father-in-law, had 29 years of experience in the mines. Foy lived near his native Berlin, where he had dropped out of high school. He served in the U.S. Army in Vietnam and worked laying brick before going into coal. He'd been a miner practically the entire time he'd been married to Denise, a Confluence girl he'd met one night at Somerset's Summit Diner.

August 5 would be the Foys' 30th anniversary, and August 7 would be

V.W.H. CAMPBELL JR./POST-GAZETTE

SAFE AT HOME: Harry Blaine Mayhugh Jr., 31, known to his family as Blaine and to his mining coworkers as "Stinky," wraps his arms around his two children, Kelsey, seven, and Tyler, eight, as relatives visit him at his Meyersdale home. When things were bleakest for Mayhugh, his father-in-law, Thomas Foy, 52, and the seven other miners trapped 245 feet underground in Quecreek Mine, he came up with the idea to write loved ones a final good-bye. Joining in his homecoming are his father, Harry Blaine Mayhugh Sr., far left, his mother, Margie, and his nephew, U.S. Marine David Mayhugh. Missing is Mayhugh's wife, Leslie, who at the time was with her father, who hadn't yet been released from Conemaugh Memorial Medical Center.

V.W.H. CAMPBELL JR./POST-GAZETTE

THE FIRST HERO: As the rush of water blasted into the Quecreek Mine, Dennis Hall Jr. raced to the phone and warned nine other men to get out, giving them time to escape even as his own life was in peril. One of those men would later ask Hall why he did it. "I had to. I had to. I had to," Hall said.

the Mayhughs' 10th, so the two couples were looking forward to taking a short trip together for a long weekend beginning August 9.

Leslie was the first of the Foys' four daughters, all of whom now are in their twenties and living nearby. The good-natured Foy dotes on them all, and especially on his seven grandchildren, including two sets of twins. Still, he is likely to tell everyone to clear out when a NASCAR race comes on TV, because he loves car racing.

Foy stands just under five feet tall, and that's how he got the name used by all his mine coworkers and friends: "Tucker," as in the nursery rhyme "Little Tommy Tucker sings for his supper."

Foy, an avid hunter famous among family and neighbors for his deer bologna, was trying to lose some weight. Four years back he'd suffered a heart attack and had angioplasty to open clogged vessels.

On his way to work that afternoon he drove his red pickup truck to the take-out window at the East End Tavern, where his wife works as a cook.

He'd always stop by or call. That afternoon she leaned out the window and asked if she should bring him home anything to eat that night, and he said no.

"I'm going to work now, hon," he told her.

She said, "I'll see you later."

The guys called **John Unger** "Ung." He lived just 12 miles down the road from the mine on 80 secluded acres folded into the hills along a meandering two-lane road. Just over the rise from the family farm where he grew up, the ruddy, round-faced 52-year-old raises corn, oats, hay, and 30 head of beef cattle. He and his wife, Sue, also raised a pair of children here, both grown and moved

V.W.H. CAMPBELL JR./POST-GAZETTE

REJOICING AT SECOND CHANCES: Rescued miner John Phillippi pauses for a moment to collect his thoughts while recalling his swirling emotions in the hours before he was hoisted from the mine. "It's just a real tough thing. You're thinking about a lot of things you could have lost, everyday things you took for granted."

away now—one to Atlanta, one to the outskirts of Butler, Pennsylvania.

Unger has been a miner for 28 years, and during that time has worked for seven coal companies. He likes weeks like that one was, when he can carve out a regular weekend. On daylight shift he and his crew would work six days. But the every-other-week swing shift of 3:00 to 11:00 P.M. meant the crew would have both Saturday and Sunday off.

That afternoon, Unger savored the prospect as he talked with a friend about going out Saturday to get his deer-hunting license. They looked over a rifle to which the friend has just attached a new scope.

But it was getting toward 3 o'clock. "Later," Unger told the friend, and hit the road.

After knocking around various laborer jobs, 36-year-old **John Phillippi**, known as "Flathead," had been happy to enter the mines in 1990 because it was the best paying job he could find, earning him about $40,000 a year with overtime. He found the work inter-

MATT FREED/POST-GAZETTE

REFLECTING ON HIS TERROR: After flood waters separated him from his companions, miner Mark Popernack thought he would die alone. But the other miners saved him, using a motorized scooping machine to haul him back across the raging water. "We constantly prayed," he said.

esting and his coworkers to be kindred souls. Besides, he was the son of a retired miner.

He lived in a bucolic section of Boswell in a blue, barn-shaped home with his wife, Melisa, and 12-year-old son. That afternoon he shared a lunch of pizza dogs—hot dogs with pizza sauce—with his son. Then, he jumped into his car at 2:20 to head for work.

Robert Pugh was "Boogie," a nickname his grandfather gave him. The 50-year-old lived by himself in a red brick house in Boswell, having divorced 11 years ago. That marriage gave him three now-grown children: two daughters living in North Carolina—one a teacher and one training to be a psychologist— and a son in Maryland who works as a golf pro. His longtime girlfriend, Cindy Thomas, lives five miles away.

A wrestling champion and all-county football center for Jenner Boswell High School, Pugh went into the mines—like his dad, like his brother—in 1970, the year he graduated. He considered everyone on his crew to be a friend, and often hung out with some of them outside the mine. Two of his favorite things in the world involved hunting: turkeys and ginseng roots.

On that day, Pugh got up around 8:00 A.M. and ripped up the carpet in his living room so he could replace it. Then he ate a TV dinner and left for the mine.

On the crew, **Mark Popernack** was "Moe." A thin 41-year-old with a brushed-back thatch of black hair, he was a 21-year veteran of the mines. He lived with his wife and two sons, ages nine and ten, in a single-story house on an isolated piece of ground past the northwest edges of Somerset, with a well-tended sweep of yard on one side, a reservoir dubbed Troll Lake on the other, and serenity all around.

On that afternoon, Popernack tinkered around some outside, careful not to tap out energy he'd need for his eight-hour shift in the mine.

Another crew member, 49-year-old **Ron Hileman**, was a 26-year veteran of the mines whom his coworkers called "Hound Dog." Lean, with a hefty, well-groomed gray mustache, he is the father of three grown children, and his wife runs a day care center in their home in the village of Gray. Right before leaving for work that day, he grabbed a jug of water.

Almost 49, **Dennis Hall** started mining at 19 and earned his nickname, "Harpo," a decade-and-a-half ago when he sported longish, curly hair. He knew the job was dangerous: A quarter-century ago, a cave-in trapped him for an hour in a northern Somerset County mine. On one job a drill let loose and walloped him in the jaw, breaking his lower denture into six pieces and gashing his face.

That afternoon he was in the mobile home park on the southern fringe of Johnstown where he lives with his wife of 23 years and his two sons.

She packed his lunch pail with a corned beef sandwich.

The crew chief, or section foreman for this crew, was 44-year-old **Randy Fogle**, who lived with his wife, Annette, and two of their three children outside the village of Garrett on unpaved Fogletown Road. As you might expect, a lot of Fogles live along it.

The son and grandson of miners, he'd been around or in mines for most of his life, working at everything from miner to foreman to mine superintendent. He had additional training as well, as an emergency medical technician.

He loved mining coal. And he really liked this crew for being efficient, smart, and flexible.

Not that there wasn't stress. He'd suffered from chronic heartburn for years.

He didn't have a nickname. His crew just called him "boss."

The crew that night was supposed to number 10. Fogle said plans had called for them to be joined by their newest colleague, 22-year-old Roger Shaffer of Hollsopple, who'd been working with them for a few months.

However, Shaffer and his wife, Lacey, had acquired tickets to attend Ozzfest at the Post-Gazette Pavilion on

MARTHA RIAL/POST-GAZETTE

PRAYERS ANSWERED: Mary Unger, 87, prayed at her Hollsopple, Somerset County, home for the safe rescue of her son, John, 52, right, and his eight fellow miners from the flooded mine. The waiting, she said at the time "was awful," particularly when rescue efforts kept hitting snags.

Randy Fogle, 44

Harry Blaine Mayhugh, 31

Thomas Foy, 52

John Unger, 52

John Phillippi, 36

Ronald Hileman, 49

Dennis Hall, 48

Robert Pugh Jr., 50

Mark Popernack, 41

July 7. But because Sharon Osbourne, Ozzy Osbourne's wife, had to undergo cancer surgery, the tour was postponed until July 24. So Shaffer took the night off and they went to the concert.

The nine members of the crew converged at Quecreek Mine around 2:30 P.M.

In the trailer where they would shower at the end of their shift, they changed into their mining gear: thermal underwear, flannel shirts, blue overalls, rubber steel-toed boots, maybe a rain coat or rain pants or both as an extra layer against the dampness. The last things they pulled on were their knee pads and their miners' helmets, which had detachable lights that they could hook on their belts.

At about 2:45, the nine went outside and exchanged news with the departing day shift—the usual chitchat about mine conditions and machinery.

One of the day-shift guys tossed in the usual see-you-later: "Have a good one, man."

Then, right at 3:00 P.M., the nine climbed onto the mantrip, a low, battery-powered rail cart, for the half-hour ride to the coal seam they were working. It was 8,000 feet, or about 1½ miles, from the mine's portal.

It didn't take the four-foot-high mine shafts to make these guys feel close. After all, they sometimes saw more of each other than they did their own families. But inside the mine they didn't always work shoulder to shoulder. Sometimes they only passed each other as they worked different parts of the coal cuts.

On that day Hileman and Unger worked together as one team, bolting the newly created mine roof to secure it so it wouldn't collapse. Fogle and Foy were the other bolting team. Pugh and Hall were car men, cleaning up debris. Mayhugh operated the scooper, a motorized vehicle with a bucket for picking up the mined coal and dumping it on a conveyor belt for transport out of the mine.

For half of the shift Phillippi operated the remote-controlled continuous miner, a low-slung machine fronted by an 11½-foot-wide cylinder with 100 teeth that grinds into the seam and chews out the coal.

Then Popernack took over operation of the miner.

CHAPTER TWO

SOMERSET AND COAL

Popernack was carrying on a tradition that has long been a part of Somerset County, where rolling hills and grassy glades have proved both a blessing and a curse.

The topography of the county's 1,000-plus square miles has enabled the easy extraction of coal from abundantly rich seams, spawning coal patch communities like the cluster of tidy homes called Quecreek and company towns like nearby Windber, a planned community built by coal barons in 1897.

But such success has come with a price.

During the 19th and early 20th centuries, miners bent their backs beneath low mine ceilings, sometimes on their knees in sludgy water, sometimes on their bellies or backs, digging with picks and shovels into the black veins crisscrossing the county.

Sons joined fathers in the mines. It was common for generations of men to earn their livings—and, occasionally, their deaths—in the mines' dark confines.

An explosion and fire in 1915 at the Orenda mine in Boswell killed 197 miners, still the state's deadliest mining accident and one of the country's worst.

Mining today is generally much safer, and has become highly mechanized work that pays good wages for rural Somerset County. Small crews with sophisticated machinery have replaced the hordes of men valued for their muscle and stamina.

It remains a familial legacy, a job passed from father to son as a way to earn an above-average wage below the ground. When men talk about their mining experience, they'll often say, "I have 10 years in the mines," owning the experience as far more than a job. That closeness with the mines—with the earth—is manifested in the tiny, tight-knit, faith-based communities spattered across the county.

Coal was one of humanity's earliest sources of heat and light. The Chinese were known to have dug it more than 3,000 years ago. French explorers discovered it along the Illinois River in 1679. The first commercial mining on this part of the continent occurred in Richmond, Virginia, 71 years later. Somerset County coal mines date to the early 1800s.

But coal's origins go back much further.

Coal is the remnant of vegetation that grew 400 million years ago in large swamps that no longer exist. The fossil fuel is often called "buried sunshine" because the trees and plants that formed coal captured the sun's energy through photosynthesis.

As layers of flora and trees accumulated, they formed a soggy, dense material called peat. Over time, as the earth's crust shifted, deposits of sand, clay, and other mineral matter buried the peat. Pressure squeezed water from the peat and the earth's heat forged chemical elements that resulted in the black combustible mineral known as coal. It's estimated that about three to seven feet of compacted plant matter were required to form one foot of bituminous coal.

Carbon is what gives coal most of its energy, and it's the reason that coal was the country's most important fuel from 1850 to 1950. As America flexed its industrial might, coal was used to melt glass, heat forges, kiln lime and cement, and process wood pulp. Coal-powered railroads moved the country's goods and coal-run steam engines drove factory machines.

In 1920, when coal was king, a miner could make 50 cents for every ton of coal he produced, and it was common for some miners to produce up to 10 tons per day. The buying power of their $1,300 annual salary would translate today to about $12,800.

Now the average mining wage in Somerset County is $33,798 a year, well above the county's average earn-

STAYING TOGETHER: Brett Stanga, nine, of Sipesville, Pennsylvania, pedals past his town's volunteer fire hall, which became a central player in a mine drama that galvanized the community and the world.

MARTHA RIAL/POST-GAZETTE

THANKS TO GOD
AND THE
RESCUE WORKERS
9 FOR

FRANKA BRUNS/POST-GAZETTE

A WAY OF LIFE: Quecreek miners Don Lydic, left, and Tony Niemic, right, personify Western Pennsylvania's tradition of coal mining; hard-working and down to earth.

ings of $23,153, or a minimum-wage job generating $10,712 a year.

But mining jobs are no longer easy to find.

In the fifties, mining and steel provided about 40 percent of the jobs in Somerset and adjoining Cambria County, where Johnstown's steel mills consumed prodigious amounts of metallurgical coal. Entire communities depended on the coal mines.

Those days are gone.

Today the biggest employers in Somerset County include health care providers; two state prisons; the Seven Springs resort; Fleetwood Folding Trailers, which makes Coleman recreational campers; and Gilmore Manufacturing Company, which makes lawn and garden implements.

It is a sign of the changing fortunes of coal that the Berwind-White Coal Company headquarters in Windber is now a museum, the Windber Coal Heritage Center, and that the largest coal producer in the county, PBS Coals, employs fewer than 400 people. Windber, which was once home to more than 13,000, now has only 4,800 residents.

PBS Coals, a privately held nonunion company that is a unit of Mincorp, controls the mining rights

JOHN HELLER/POST-GAZETTE

at Quecreek, although the actual digging is done by Black Wolf Coal Company, an independently chartered company whose chief executive is David Rebuck, a miner who was once president of a PBS sister company, Rox Coal.

The complicated mix of companies doing business with each other is common in the coal industry to avoid unionization and to lower costs such as insurance coverage.

Today roughly 1,000 of Somerset County's 80,000 residents work in mining at fewer than 40 mining sites. Most are surface mines that strip away the earth to reach the coal; a few are deep mines like Quecreek.

Three shifts of miners show up daily at Quecreek, lunch pails in hand, to descend more than 200 feet underground for their eight-hour tours of duty. The 18 men on the mine's 3:00 P.M. to 11:00 P.M. shift had done just that on Wednesday, July 24, riding into the damp darkness.

Now they were deep in the mine, chewing up the coal.

CHAPTER THREE

THE BREACH

Routine turned to mayhem in two seconds flat.

About 245 feet underground, Popernack was running the continuous miner machine, the steel teeth of its 11½-foot-wide cylinder spinning and chewing into walls of coal. He was working near the sixth of seven pathways cleared into the mine.

He knew it was about 8:45 at night, two hours from the end of a shift that seemed like any other.

But five minutes later the machine chipped straight through a wall that was supposed to be hundreds of feet thick. It had broken into the adjacent, abandoned Saxman Mine No. 2, unleashing more than 150 million gallons of groundwater into Quecreek.

"Everybody out!" screamed Foy, fear in his voice. "We hit an old section! There's a lot of water!"

Inaccurate maps had falsely led the men to believe they weren't anywhere near the old mine. In the half century since it had been emptied of coal, the Saxman Mine had become a subterranean reservoir.

Hileman and Unger were about 100 feet away, up a crosscut. Their machinery and ear protection covered the roar of the torrent and Foy's scream. But then they

DRAINING THE PIT: At the entrance to the Quecreek Mine, workers raced to pump out more than 18 feet of water accumulated in the pit. It was just one of several sites, including a cornfield, where pumps of all shapes and sizes were set up to suck the water out as part of the rescue. Joe Gallo, the first engineer called by Black Wolf Coal Company, said there was no time to be particular. He told one man to bring every pump available no matter the size. "I didn't care where he got them. We had an emergency at the mine. Call your distributors or your suppliers and tell them to get every conceivable pump. Diesel pumps, I don't care what they are. Underground submersible pumps. I was in a daze. I didn't have time to think about what I was doing. We called pump suppliers. We called drillers. We called other personnel in the company to get them to the site. Electricians, mechanics, welders."

DAVID SWANSON/PHILADELPHIA INQUIRER

Setting the scene

Nine men escape, nine men are trapped and rescuers must work aggressively yet cautiously to avert tragedy.

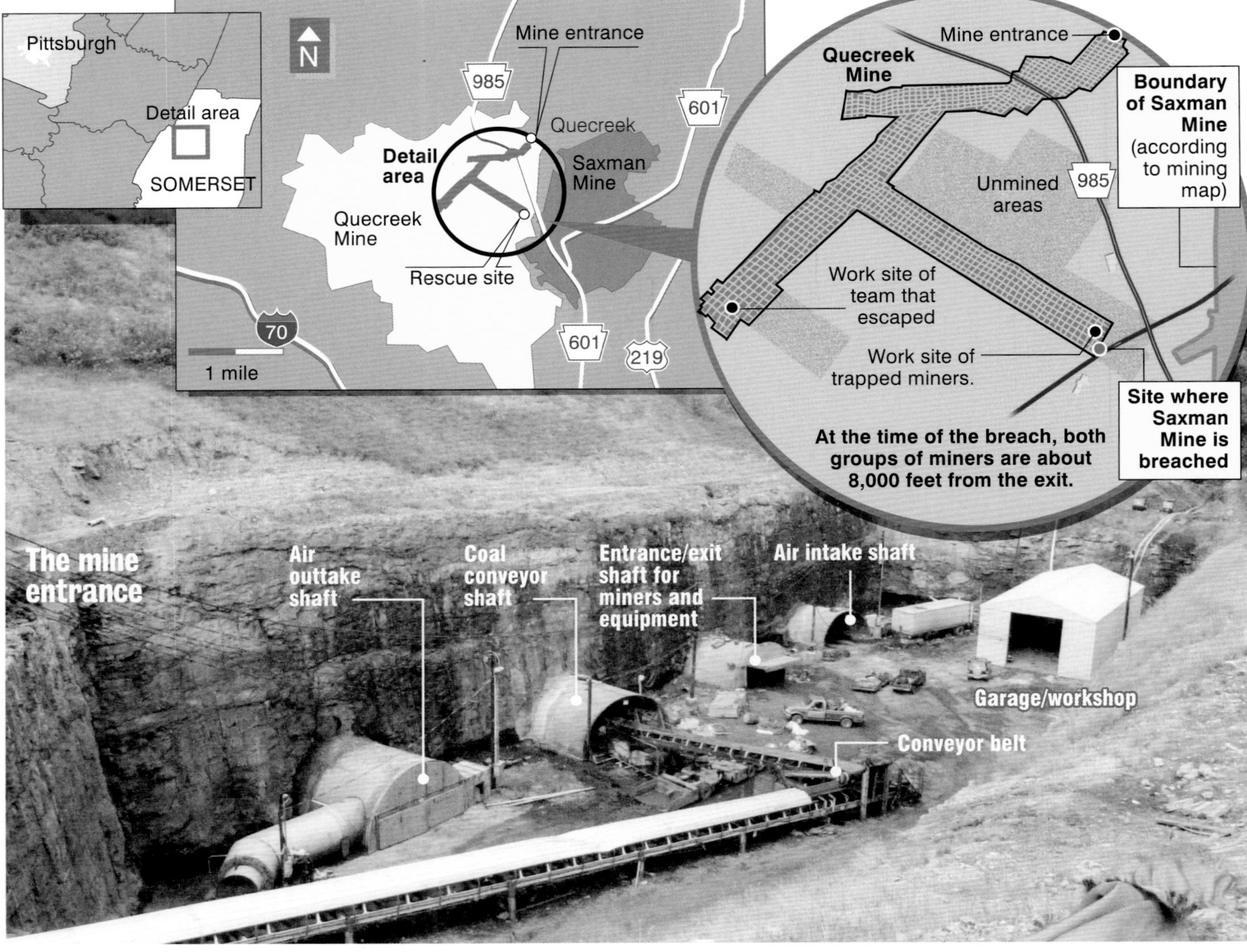

GRAPHIC: STACY INNERST, STEVE THOMAS AND JAMES HILSTON/POST-GAZETTE PHOTO: DARRELL SAPP/POST-GAZETTE

DARRELL SAPP/POST-GAZETTE

THE BIG QUESTION: The old map of the Saxman mine is compared to the new map of Quecreek. The maps indicate that Quecreek miners should have been hundreds of feet away from Saxman, but one bite of a continuous mining machine put them through the wall and facing a raging river of Saxman water.

saw the water.

In a second, maybe two, water that had been gushing from the hole exploded through the wall, washing over Popernack's miner machine and carrying off another five-ton piece of equipment called a load-center.

"Harpo! Get the hell out! Get out now! Get out now!" Popernack screamed to Hall, who was right behind him in an 18-foot, electric-powered shuttle car loaded with the coal Popernack's miner had cut from the walls.

Hall heard the warning and seconds later drove away, maybe 210 feet through the coal cuts, before the power feeding the car was knocked out by the water.

Popernack jumped away from the mining machine, saving his life momentarily, but nearly dooming himself to die alone.

The instant flood created a furious river, cutting off Popernack from his coworkers and spoiling to suck him under and whisk him away if he stepped into it. The orange water roared, drowning out shouts among the men.

In a space barely four feet high and with only the light from their headlamps to guide them, the miners on the other side of the torrent began their desperate

Facing the problems

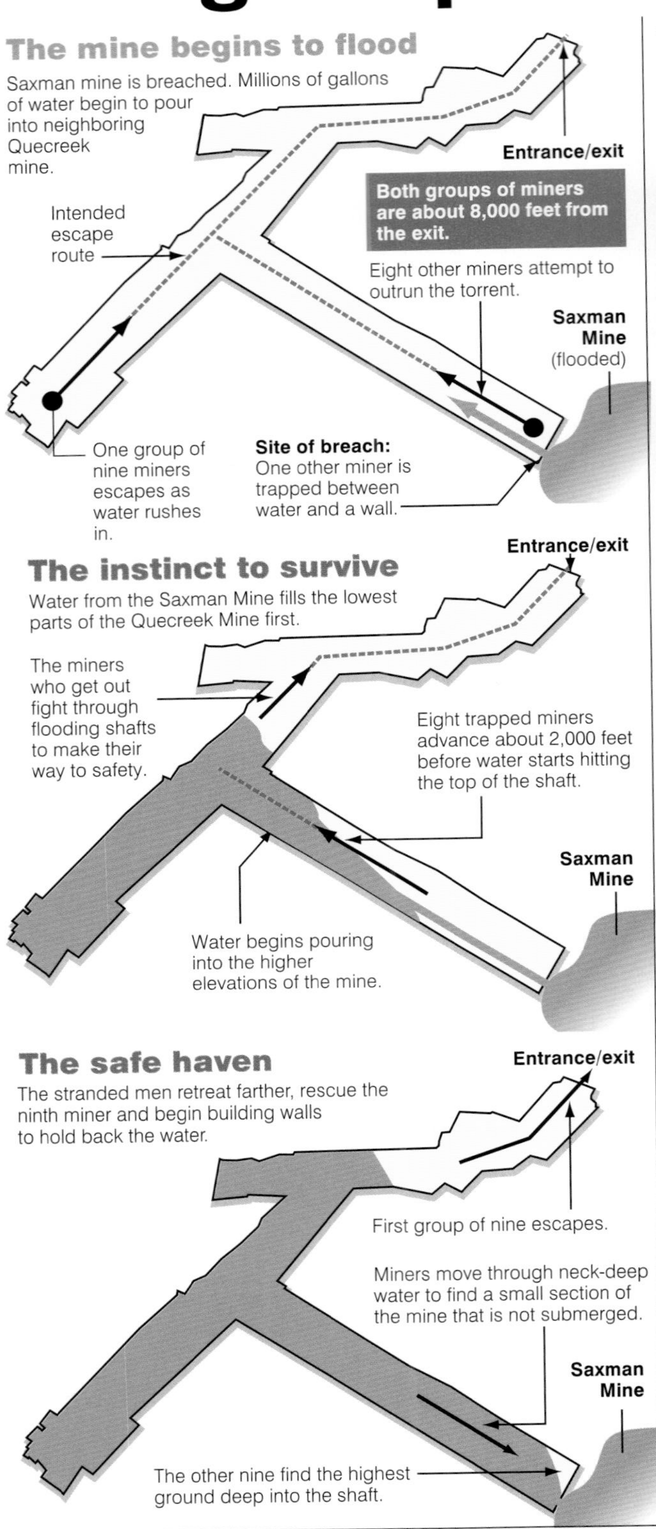

The situation

Nine miners are trapped 245 feet underground in a flooded shaft, but breathing because air is being pumped to them.

Graphics not to scale. Schematic only.

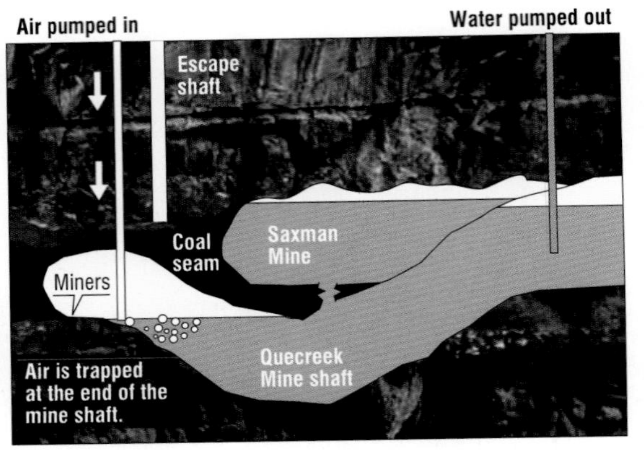

The peril

If a "super drill" trying to reach the men enters the shaft too rapidly, the miners' life-sustaining air pocket could be destroyed.

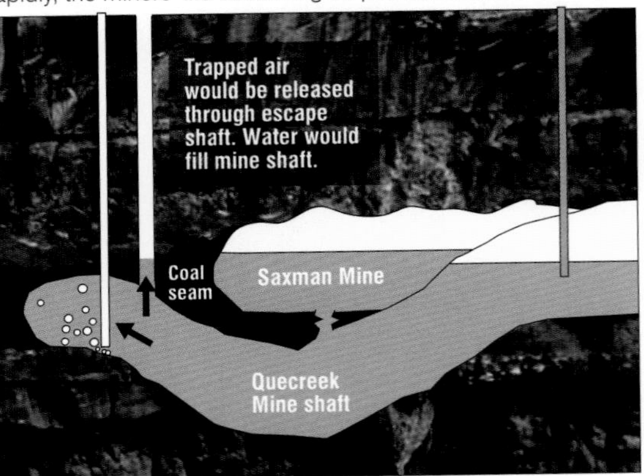

The solution

Rescuers pump water from the mine. Water recedes and pressure is relieved as air pocket expands . The drill can now continue without fear of flooding the shaft.

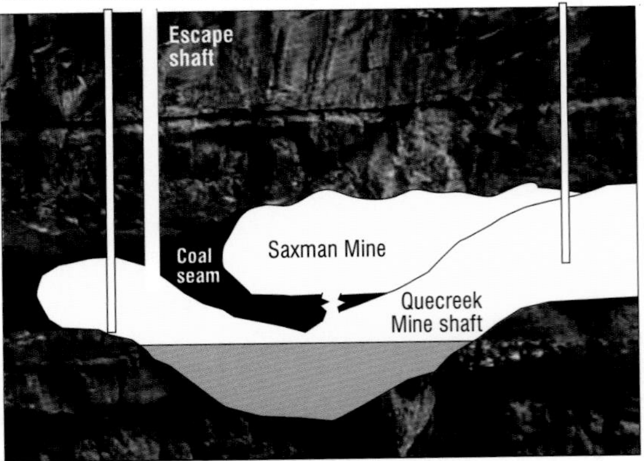

Sources: Pa. Dept. of Environmental Protection, Christopher J. Bise/Penn State professor of mining and geoenvironmental engineering, accounts of miners

James Hilston, Stacy Innerst, Steve Thomas/Post-Gazette

MARTHA RIAL/POST-GAZETTE

A NARROW ESCAPE: Quecreek roof bolter Doug Custer of Berlin, Somerset County, recounts his successful escape from a section of the flooding mine. Custer was part of a second crew of nine workers, all of whom managed to get past the torrent that trapped the others, because they received a telephone warning from miner Dennis Hall. Hall didn't get out until 78 hours later.

race to get out.

The water thundered down Entry No. 6, then broke off, filling other entries and cross cuttings.

Amid the fury came the first move that saved lives. Foy and Fogle yelled at Hall, who was nearest the mine phone, to call the second crew of nine in another part of the mine.

Having ditched the shuttle car, Hall ran about 70 feet to the mine phone and tried to warn those working in a lower section farther away from the break.

Nobody answered. Maybe the machinery was too loud. Maybe nobody saw the flashing light by the phone. Forty seconds seemed to stretch into forever until Hall heard a voice, but not from the mine. The miner who was stationed on the surface for emergency purposes had picked up the phone. Hall was telling him what had happened, when one of the second crew also picked up. To him, Hall screamed:

V.W.H. CAMPBELL JR./POST-GAZETTE

THE TENSE WAIT: Jeffrey Stanchek, a mine rescue instructor from the state Bureau of Deep Mine Safety, wipes his eyes while waiting with other rescuers for the super drill to arrive so they can start boring a rescue shaft. Rescuers had already drilled a six-inch hole and were pumping warm air into the mine, but weren't sure if the miners were alive or dead.

"Get out! Get out, goddamn it, get out! You got major water!"

Word of the water spread quickly through the second crew. They were roughly 3,000 to 5,000 feet away from Hall's group.

"Shut everything down," a crew member yelled to miner Doug Custer, 45, of Berlin in Somerset County. "We're getting out of here!"

Joe Kostyk, 42, of Salix in Cambria County, was running the continuous miner when the call came in.

He, Custer, and three others immediately piled into the mantrip and started driving out the common entry tunnel, picking up a sixth crew member along the way. Two other crew members were two minutes ahead in a golf cart. Another was somewhere else in the mine.

The men in the mantrip met water within five minutes. It sloshed up over the sides of the buggy, which the men had to abandon because the water was already too deep at a foot-and-a-half. The guys in the golf cart had to ditch it, too.

They escaped through a door down into another passageway. Now on foot, Custer and his coworkers kept looking at the cement-block walls, trying to gauge the water's height. When they saw water coming up over of the top of the blocks, they knew the water was at least four feet high in the adjacent entry tunnel.

When they made their way over into the Entry No. 6 tunnel, a swift, cold current was there to meet them, knocking them down as if they were rag dolls. The water, about 20 feet across, was only a little over 2 feet high, but for men walking crouched in a cramped, 4-

foot-high space, it was up to their chests. They had to get across it.

Custer, a roof bolter, thought he was about to die. For all they knew death lay ahead even if they made it across. But they would have met certain death had they not traversed the torrent.

Three crew members made it across ahead of them. Custer and Kostyk made their way through the channel. Then they yelled back for the rest to follow. When the other four made it over, Kostyk felt a bit better because they were heading uphill, to higher ground.

And at that point it was dry. They walked back over to Entry No. 5 tunnel and met up with two members of their crew who had gotten out ahead of them and were driving a golf cart to freedom. The six jumped onto the cart—two on the front fender, two in the middle, and two in the back—and all eight rode out of the mine. The ninth crew member was already out, trying to reach Hall's crew by mine phone.

The men were clear of the water within 15 to 20 minutes, but because of the distance it took them 45 minutes to get completely out of the mine.

Once outside they notified the mine owner and mine foreman, but waved off medical treatment for themselves, all the while looking back for lights or any signs of the first crew.

There were none.

Back at the breach the first crew had begun its attempt to escape only after realizing they couldn't take Popernack with them. He screamed at them to save themselves. The roar of the water swallowed his words.

In the dark, Popernack shook his head, and the beam from his helmet light swept back and forth. He didn't think he could make it. He figured he'd probably die there.

Using their safety training, the other eight converged at Entry No. 4, at the conveyor belt that moved the mined coal out of the shaft. They prayed this was their way out.

The rush down the passageway toward the mine mouth was at full bore. The men raced through the water, grabbing the coal conveyor to keep from falling, from going down in water too violent to ever let them get to their feet.

In the passageway, four-and-a-half-feet high in places, they ran as fast as stooped men could run. But because they were hunkered down as they ran, the water—now three to three-and-a-half-feet deep—was reaching their necks.

Unger wasn't sure how long he could continue the pace. He thought he was on the verge of a heart attack..

When the water got to be too much they hopped on the conveyor—which had stopped when the water cut its power—and crawled along it. They scrambled across coal on the conveyor, falling, bloodying their knuckles, and realizing that the water blocking their way was rising toward them.

As Hall rushed to the conveyor and joined the race to get out, the water nearly took him under. Fogle was running right behind him.

When Hall tried to clamber onto the conveyor but couldn't, Fogle, at 200-plus pounds, grabbed all 180 pounds of Hall by the coveralls and tossed him aboard the stilled conveyor.

Somewhere in the din, Hileman called out, "Can we make it?"

The two youngest, Phillippi and Mayhugh, had led the struggle down the conveyor for about 2,000 feet.

As they looked forward their helmet lamps illuminated their worst fear: 100 feet ahead, the water had gotten so high it was hitting the roof. The water had outrun them down entries 5, 6, and 7 and was now enveloping them.

The two young miners yelled at their fellow crew members to turn back.

The water, now at chin level, was so overpowering that the men had to pull on the conveyor as they fought to reach higher ground.

And at one point Foy turned to his son-in-law, Mayhugh, with the saddest eyes the younger man had ever seen.

"We're in trouble," the veteran miner told him.

"I know, Tom," Mayhugh responded. "I'm too young to die. I'm not afraid to but I got two little kids. This ain't the way for us to go."

CHAPTER FOUR

THE RESCUE BEGINS

Dave Rebuck, the owner of Black Wolf Coal Company, which mines Quecreek, was getting ready for bed just before 9:00 P.M. Wednesday when his wife Annette handed him the phone.

It was the mine's "outside guy," the miner who stays above in case something happens. Something had.

A breached wall. A flooding mine. Men possibly trapped.

Rebuck drove the 30 minutes to the mine and called state and federal officials. That set emergency workers in motion, including the state's deep-mining safety expert, Joe Sbaffoni.

Sbaffoni loves miners. Tough as a boot and straight as a string, they are the best part of his job with the Pennsylvania Department of Environmental Protection.

Sbaffoni, 51, has 32 years of experience in coal mining, the last 18 with state government. None of it prepared him for the call about the Quecreek flood.

He was sitting at his kitchen table in Fairchance in Fayette County about 9:30 P.M. Wednesday when his district mine manager, Lynn Jamison, telephoned. Nine men were missing.

"My heart stopped," Sbaffoni said.

He hurried to Uniontown for a meeting with Richard

ON THE MARK: In the Quecreek Mine rescue, traditional techniques and *Star Wars* technology guided the effort. After engineers at the rescue command center plotted locations for the air shaft, the rescue shafts, and the pumping stations, technicians like Robert T. Long used a Global Positioning System or GPS to pinpoint the drilling sites. The GPS takes signals from a satellite to locate longitude and latitude. Long is employed by Somerset-based CME Engineering. V.W.H. CAMPBELL JR./POST-GAZETTE

GEHL
4610

KOBELCO

Stickler, director of the state's Bureau of Deep Mine Safety, and mining engineers Bill Bookshar and Tom McKnight. They studied maps and then headed for Quecreek to start what they hoped would be a rescue mission.

At 9:53 P.M., Somerset County's 911 center got a call requesting an ambulance at the mine. Soon after, word about the trapped men reached law officers.

The job of telephoning the families of missing miners fell to State Police Corporal Robert Barnes Jr.

He was brief and businesslike. An "incident" had happened in the mine, he said, volunteering no details. He asked the families to come to the Sipesville fire hall, where they would be told more.

One of those calls went to Denise Foy, wife of Thomas Foy.

She was sitting at home with one of the couple's grandsons, eight-year-old Tyler Mayhugh, who'd come from her daughter Leslie Mayhugh's house in Meyersdale. The two of them were watching a movie called *Surviving the Game*.

Leslie and her parents are so close that it could have been Leslie calling when the phone rang around midnight.

It was Barnes, Denise Foy said.

"He said, 'Mrs. Foy?' And I said 'Yes.' 'Mrs. Tom Foy?' 'Yes.' 'There's been an accident at the mine.' "

The trooper said some miners had escaped and some had not, but he didn't say anything specific about her husband.

Denise immediately called Leslie Mayhugh, who reminded her mother that her husband, Blaine, was working at the same mine. That call was interrupted when the state trooper called Leslie too.

Barnes, 47, has been stationed at the Somerset Barracks for 18 years. He has seen his share of unexpected deaths, usually resulting from car crashes. In those cases he found it easier to console families when he had a clergyman at his side.

Barnes made hundreds of decisions the night Quecreek mine flooded. Perhaps none was more important than deciding to bring the miners' families together and having ministers on hand to comfort them. He called his friend and pastor, Barry Ritenour, who has two United Methodist churches in Somerset County, to ask a favor. Could he spend the night at the fire hall with the families?

By then it was 11:00 P.M. and Ritenour had just climbed into bed. He told Barnes he would make the 15-minute drive as soon as he got dressed. Barnes felt better. He had no sense of whether the miners were alive, but he knew that Ritenour would be a soothing presence.

Ritenour wanted to make certain that families did not arrive at the fire hall only to find it empty. So he called another pastor, Joseph Beer of Laurel Mountain United Church of Christ. Beer lives so close to the fire hall that he could walk there in two minutes.

Beer arrived before anyone else. He stood alone for awhile in the century-old building, which once was a school.

Squat and white, the fire hall has crumbling steps out front and a rusted metal angel propped against a sign in the front yard.

Inside, there's not much decorating the tan-paneled walls except for a few trophies (including one for "best appearing pumper" at a parade) and an artist's rendering of the famous photograph taken on September 11 of three firefighters hoisting a flag at

Previous page

PREPARING TO DRILL: With a six-inch hole dug and compressors blasting air to the trapped miners, rescuers began the arduous task of preparing to dig the first rescue hole. Joe Gallo, an engineer at the site, said: "It takes time to set the casing. It takes time to set the rig in place. It takes time to set up all the air compressors that the rig requires. It's agonizing. The time just seemed to crawl. It seemed like it took forever to do anything with those guys under there waiting for us. We knew they were alive. Seconds seemed like hours."

V.W.H. CAMPBELL JR./POST-GAZETTE

V.W.H. CAMPBELL JR./POST-GAZETTE

BATTLING FOR TIME: While workers struggled to pump out the water engulfing the mine, Joseph Sbaffoni, of the Department of Environmental Protection's Bureau of Deep Mine Safety, on the left, briefs DEP Secretary David Hess about the massive rig summoned from West Virginia to drill an escape chute for the trapped miners.

New York City's "ground zero."

But the families weren't looking for atmosphere when they began to troop in around midnight.

They seemed shell-shocked. Still, questions abounded. Answers were harder to come by.

In those first desperate hours, nobody from Black Wolf Coal Company or mine-safety agencies could speak to the most critical question: whether the men had survived the onrushing waters and found a sanctuary in the mine.

"When I first got there, I thought, 'They'll have 'em out in two hours,'" Denise Foy said. Tom had taken her into mines before, "so I pretty much knew what they were going through." She figured that since the accident happened near the end of the miners' shift, they would already have eaten their food, and that their cap light batteries would be running low.

When Ritenour, the pastor, entered the fire hall, he saw a room full of sad faces. Then he noticed one woman with a perpetual smile. She was Susan Unger,

DARRELL SAPP/POST-GAZETTE

STAYING VIGILANT: Black Wolf Coal Company spokesman John Weir, at left, and owner Dave Rebuck were at the rescue site every step of the way. Rebuck was confident the men would be brought out.

wife of miner John Unger.

As the hours stretched on, many in the fire hall would lose their composure and lash out in frustration. Susan Unger never did.

She has multiple sclerosis and needed a walker to move around the room, but she radiated optimism.

"I know they're going to come out," she told the others.

Her resolve set a tone. They had to believe.

Few reporters had yet assembled around Quecreek, but Barnes knew they would be coming in droves. He wanted the families sequestered in the fire hall for many reasons. For one, he knew police would restrict access to the building, sparing families from incessant interview requests.

Beyond that, he thought the families might draw strength from one another and the pastors, whose number would grow to eight, one for each church the nine families attended.

He decided to put the media outside Casebeer Lutheran Church, a couple of miles from the fire hall. But he soon switched the media headquarters to a parking lot at a closed Giant Eagle grocery store, which was nearer to restaurants and motels and had ample space for the hulking satellite trucks that would beam the miners' story around the world.

Barnes also sent a state trooper to the road leading to Quecreek Mine. He wanted it kept clear for emergency workers and vehicles bringing rescue equipment.

Danny Sacco, president of the Indiana County-based Special Medical Response Team, and Dr. Jim Dixon had loaded an SUV with their best guess of what might be needed immediately. The big trucks with the rest of their equipment would arrive 40 minutes behind them.

While Sacco drove, Dixon got more information from the Somerset 911 center. Having been told that there had been a mine accident, they had assumed there was a fire or cave-in causing burns and crush injuries, not flooding and hypothermia.

The Quecreek situation was dire. Sacco thought, "This is not going to be a rescue; it's going to be a body recovery."

But he also told himself, "We have to do whatever we can to give these guys the best possible shot."

As Barnes covered every base outside the mine, company executives and government mine experts were planning the rescue mission.

Their first step was to pinpoint the location of the stranded men.

Mine operator Dave Rebuck had set up a command center in his office 200 yards from the mine portal. There they had enough room to spread out maps showing the 8,000-foot Quecreek Mine's passages.

John Urosek and Kevin Stricklin of the U.S. Department of Labor's Mine Safety Health Administration said everyone agreed the men would find high ground. It was that simple: they had to keep their heads above water.

After narrowing down the location with common sense, they turned to science to make sure they would hit a central part of the mine shaft with an air pipe. Using global positioning satellites, they determined that the drill site should be on a patch of the nearby Arnold family farm.

The Arnold dairy farm sits alongside quiet Route 985, just a few miles north of Somerset. A narrow side lane, bordered by Queen Anne's lace and crown vetch, leads up a hill to the barn and the main house.

On one side of the lane is a pond, partly choked by algae, where the Arnolds' cows drink. On the other side of the road is a humble hill. It was at the foot of that hill, visible from Route 985, that the rescue site would be set up.

It was 2:00 A.M. when the rescue crews converged on the site, waking the Arnold family. At first, family members thought they were being invaded. All they could see were flashlight beams sweeping across their property. Homeowner Bill Arnold thought, "Around here, somebody out around your sheds at night is a call to arms."

Once he realized what was happening, Arnold joined the rescue effort. Lori Arnold opened the house to muddied workers, giving them a place to rest, freshen up, and grab a bite to eat.

In a strange way, the rescuers felt confident. They believed they had found the miners. Now they had to set them free.

The first line they planned to drill to the miners would funnel fresh, 190-degree air to the men. It might also offset some effects of the cold water the miners were sitting in. Water temperature was estimated at 55 degrees.

Because the mine shafts are so low, the miners could not stand. Rescuers feared they were hunched in water to their chests or higher, a circumstance that could drop body temperatures to dangerous levels below 95 degrees.

Along with getting fresh air to the men, they knew

LAKE FONG/POST-GAZETTE

URGENT CARGO: Pennsylvania State Police escort a convoy of trucks carrying sections of the super drill over Route 601 to a field above the mine. There, the drill would be assembled and set to work piercing the earth and rock to create an escape tunnel for the miners.

they would have to bore other holes to pump water out of the mine. Black Wolf Coal Company workers began waking up anybody with a pump and hunting for drills capable of boring 300 feet into the ground.

Workers at Somerset's 911 center joined in the search. They realized it would be fruitless to call drilling companies, all of which had long ago closed for the night. So they improvised.

Dispatcher Jeremy Coughenour, 28, remembered that his old Sunday school teacher, Judy Bird, ran a drilling business. He called her at her home. Bird, her husband, Paul, and their three daughters, ages 16, 18, and 21, hustled to the mine, where they would set up a drill to bore one of the holes for water removal.

The critical job of drilling the six-inch air hole went to a company owned by Louis Bartles of Somerset. His four-member team started working about 3:15 A.M. Thursday.

That drill cracked through the mine shaft one hour and 45 minutes later, a blistering clip for obliterating what turned out to be 240 feet of rock.

Alex Nicoletti, 33, ran the drill with the right mix of urgency and restraint. Had he pushed the pace any harder, he could have destroyed the drill and the rescue would have stumbled before it could start.

With the air pipe inserted in the mine, rescue workers tapped on it. Minutes passed. Then, in answer, they received nine strong bangs on the pipe. Could it be? Was this tapping at 5:15 A.M. a signal that all nine were alive, or had one of the miners coincidentally smacked the pipe nine times? More tapping back and forth. But the signals were inconclusive.

Nobody above ground knew for sure, but hopes soared. They were almost certain all nine were alive.

Knowing that someone was alive made it paramount to stop the rising water, which monitors showed was approaching 1,825 feet above sea level. At that rate rescuers knew they had perhaps an hour before the area where the miners had taken refuge would be under water.

Sbaffoni saw another problem. A blast of "bad air," containing low oxygen levels, had shot out of the mine when the drill broke through. He worried that, before long, the miners might have trouble breathing.

Urosek, a mine ventilation expert, saw this as an opening to raise an idea. Air pockets sometimes exist in flooded caves. Why not create an air pocket in the mine? That would mean pumping compressed air through the bore hole to press against the water and prevent it from rising.

Urosek's plan had never been tested in the United States, but he urged the others to back him.

"Nobody was screaming or yelling, or saying 'John, that's just stupid,'" Stricklin said. "But there was some skepticism."

State mine engineer McKnight thought Urosek's idea would work. He whipped out a calculator to determine how many pounds per square inch of pressure would be necessary to keep water out of the miners' haven.

They ordered the hole sealed. Then they told the drill operator to crank up his rig's air compressor, which could maintain 100 pounds per square inch at about 1,000 cubic feet per minute.

The amount was excessive, but rescuers were not sure if they could close the hole tightly. Better to have plenty of pressure, knowing that some of the air would leak.

Shortly before 6:00 A.M., Sbaffoni explained the plan to workers at the drill site. Volunteer firefighters sealed the hole around the air compressor pipe by inserting sturdy bags normally used to lift wrecked vehicles off people. The experts and those doing the dirty work were becoming a team.

Now Governor Mark Schweiker was about to become its leader.

During the next three days he would serve as the state's chief spokesman on the rescue, a job that transformed him into the nation's most recognizable governor outside of Minnesota's Jesse Ventura and Florida's Jeb Bush.

Schweiker did not intend it to happen that way until he walked into the fire hall Thursday to meet the families.

"They needed to know their governor was going to stick it out," he said. "It's my state. These are my people."

So Schweiker, who ascended from lieutenant governor 10 months ago and is not running for election this year, stepped onto a world stage for the first time.

When he faced the press Thursday afternoon, his usual business suit, starched white shirt, and spiffy necktie were nowhere to be found. Schweiker switched to jeans and chambray shirts, enabling him to crawl into the muck with rescue teams.

With the air passage in place, crews focused on preparing the ground for about 10 additional drills that would accommodate water pumps.

In the meantime, Gene D. Yost Drilling Company of Mt. Morris in Greene County talked to coal company executives about using a "super drill" to create an escape tunnel for the miners. With a 1,500-pound bit the super drill could smash through the 240 feet of stubborn earth and create a wide enough chute to bring the men up, said company executive Duane Yost.

The bit was in Greene County and the drill rig was at another job in Clarksburg, West Virginia. It would take hours to haul the equipment to Somerset County.

Mine-safety workers toyed with the idea of sending divers into the mine. They might be able to organize such a daring rescue mission before the super drill could get cranked up. But the danger of swimming through 1.6 miles of flooded shafts seemed overwhelming, even for the best frogmen. And then what would they do with the miners once they reached them?

So hope was pinned on a super drill, which most of the rescuers had never even seen before.

BILL WADE/POST-GAZETTE

POINT MAN: Governor Mark Schweiker, right, and David Lauriski, U.S. Assistant Secretary of Labor for Mine Safety Health Administration, assess the steady progress made by the super drill as it chews through the earth above the mine. Schweiker's political profile rose nationally and his leadership drew widespread praise during the crisis.

CHAPTER FIVE

DESPERATE TO SURVIVE

Before the miners were able to signal the surface that they were alive, they had a fierce struggle ahead of them.

Just after they had reversed direction on the conveyer belt, they managed to move back up the beltway as they headed for higher ground. Still, the water rose to their chins. Walking hunched over in the four-foot high shaft, they now had to crane their necks back to keep their faces above the water.

Every now and then they would veer off to cross shafts, but everywhere they turned there was water. At long last they got to an area high enough that they no longer had to struggle to keep their heads above water.

There Fogle tried to beat a hole with a mason's hammer through a cement block wall to gain entrance into another passageway, possibly a way out. He pounded as hard as he could until he exhausted himself. Phillippi and Mayhugh took over and ran out of gas themselves. Another miner picked up the hammer but by then the water had risen over their heads again and they had to move to even higher ground.

Still alone, Popernack rigged a hose under his armpits as a harness, wondering if he could tape hooks to his hands and somehow get across the raging river, clinging to roof bolts to steady himself. He was desperate.

Suddenly he saw a light. It was from Phillippi's helmet lamp in a passageway across the water from him. The rest of the crew returned there, too.

"Get me over with you guys!" Popernack shouted.

"I can't," Fogle told him. "The water's too fast. We'll have to wait for it to slow down."

Before long it slowed just enough for Fogle to steer a small highlift called a scooper into the torrent, slowly easing it forward into water that could easily grab it and carry it off.

"Randy, be careful about what you do," Hall told him. "The water takes it, it's gone."

With the scooper's bucket raised above the rushing water, Fogle yelled to Popernack, "Jump!" And

Nine men out

How a crew of Somerset County coal miners became trapped in a flooded shaft, then hung on for 78 hours while the world rooted for their rescue.

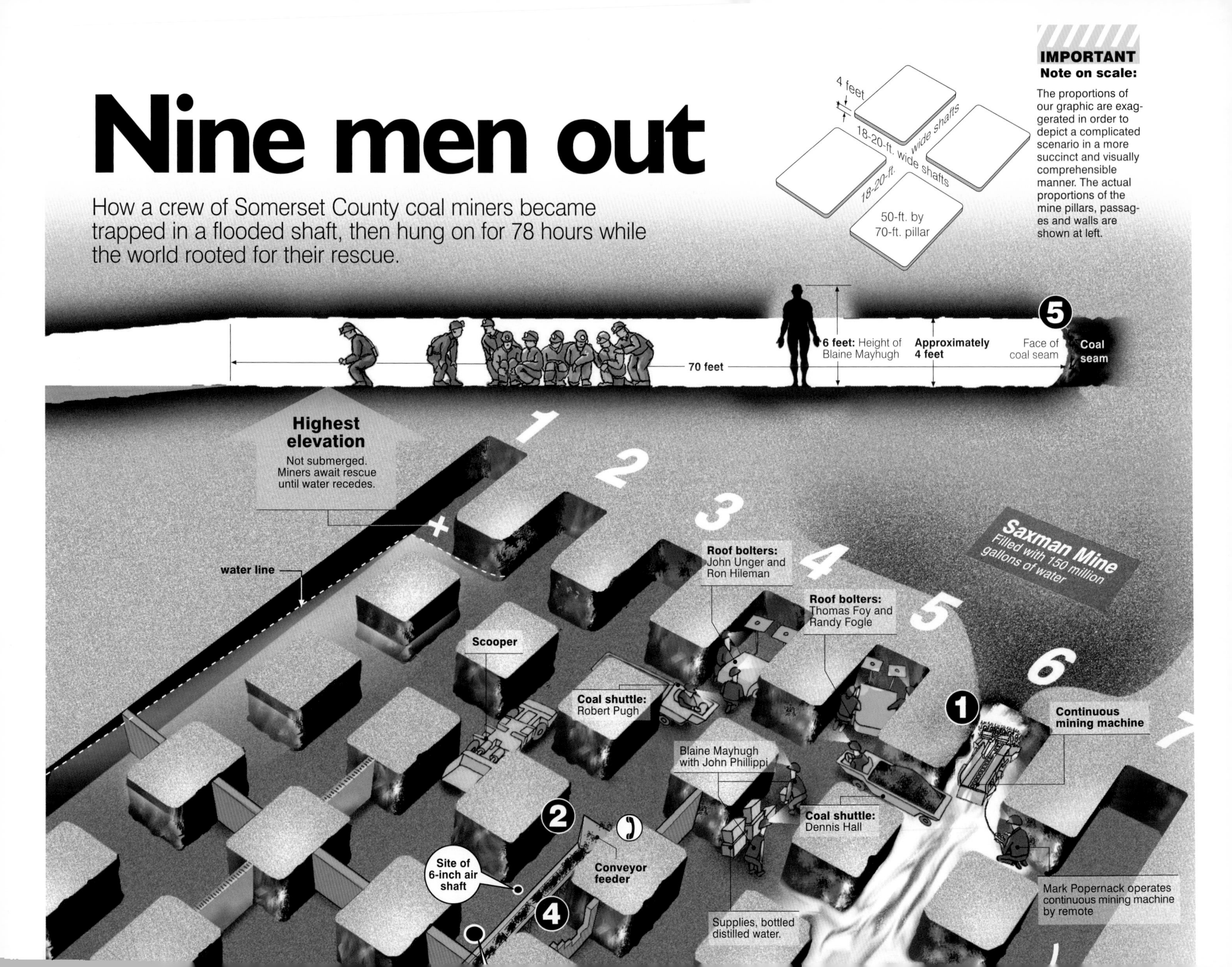

removed to construct wall.

Site of escape shaft

Entry No. 5 is miners' travel way

Conveyor belt

Cement block walls to direct air flow.

The rescue capsule
An unmanned capsule is sent into the shaft. One by one the miners are lifted to safety.

24" X 36" door slides upward

Communicating with the trapped miners
Rescuers lowered a hands-free microphone through the air shaft to the miners. Seconds later, they were talking to rescuers.

30 inch escape hole

Drill bit

245 feet to surface

6-inch pipe for air supply

Scene 1
- The wall is breached at Entry No. 6.
- Mark Popernack jumps out of way and into Entry No. 7, where he gets trapped.
- Thomas Foy shouts, telling the crew to get out.
- Dennis Hall gets to the phone and warns other miners of the breech.
- Miners head for the conveyor belthead at Entry No. 4 to follow the conveyor out.

Scene 2 (above)
- Miners advance about 2,000 feet to discover the water is "roofing," or hitting the mine shaft roof.
- The men turn around, retreat against the current.

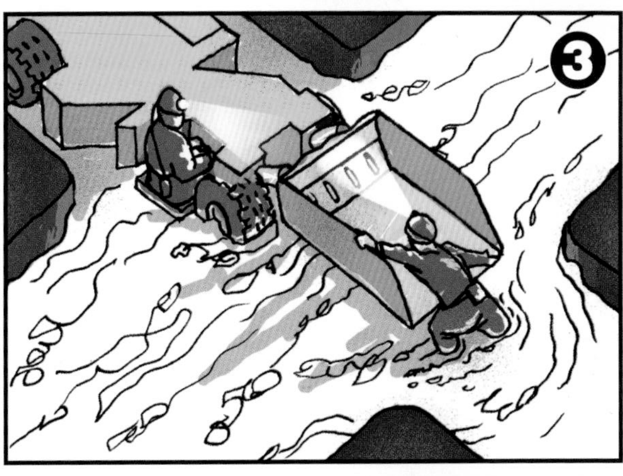

Scene 3
- Upon returning, the miners see that Popernack is trapped between entries 6 and 7.
- Randy Fogle maneuvers the scooper, a small highlift, and raises the scoop to bridge the torrent of water so Popernack can jump in.

Scene 4
- Three hours have elapsed; miners are gathered at the conveyor in Entry No. 4.
- The crew decides to build cement block barriers to try to hold the water back.
- An air shaft is drilled from above into Entry No. 4 just as breathable air is running out.
- Five walls are built. Advancing water makes it impossible to finish the sixth wall.

Scene 5 (top)
- Retreat to face of Entry No. 1
- Miners take turns banging on the pipe until water gets too high.

Graphic by Daniel Marsula and James Hilston/Post-Gazette

Popernack, the lightest of the crew, leaped into the bucket. Fogle eased the scooper backward. The nine miners were once again all together—but together in much danger.

So dire was the situation at that point that the sole advantage, it seemed, was that they would be together when they died.

But the miners figured that help was trying to get to them. They heard drilling.

Early Thursday morning, at the entries that had been dug through the earth to get at the coal, they tried building dikes with the cement blocks they normally use in their work. Their plan was to shield themselves from the water in a 100-foot by 150-foot rectangle.

It was hard, frantic work and they were breathing low-oxygen air, called black damp, that had poured in from the Saxman mine.

"Is it just me? I can't breathe," Mayhugh said. He and Unger vomited. Everyone was struggling, some so much they couldn't work anymore.

But then, at 5:10 A.M., their spirits soared when a six-inch drill cut into Entry No. 4, where the conveyor belt was located, and a pipe dropped down. They repeatedly tapped on the pipe—eventually in a sequence of nine—to indicate all were alive. And then heated, compressed air came roaring through the pipe, providing them with much needed oxygen.

The air was being pumped at such a high rate that the roar deafened them and hurt their ears. But they could put up with the discomfort; the pipe meant rescuers knew where they were.

Still, the foul-smelling water rose foot by foot toward them, eventually covering the air shaft, muffling the sound and giving them some relief. But it also prevented them from pounding on the pipe, so they began to rap on the rock ceiling further away—nine taps every 10 minutes—hoping someone using specialized listening equipment would hear them.

By noon Thursday they had completed five cinder block walls, but it was all to no avail. The water overtook them as they worked on the sixth. They had to retreat to the highest ground, about 300 feet from the airshaft, near Entry No. 1. There they would wait for what seemed the inevitable as the water rose closer and closer.

The water lapped 70 feet away. It was moving too quickly. Fogle, their leader, gave it to them straight: in another hour, he estimated, all of them would be dead.

The men knew he was right. They had to be realistic. There was quiet. There were tears. There were silent prayers.

Mayhugh pulled out a pen, grabbed a piece of cardboard from the ground, and wrote a note to his wife and kids, telling him he loved them. He dumped out drill bits from a white plastic bucket, put in the note, and offered the pen to everyone else.

Every man did the same, writing their good-byes to loved ones. When nine notes were inside the bucket, the airtight lid was snapped on and the bucket was lashed to a boulder so it would be found.

Foy then grabbed a 3/16 inch, plastic coated steel cable from the materials normally left around a working mine. He looped it onto their miners' belts, saying that if they had to die they would do so as a team, as a family. This way their bodies wouldn't be scattered through the mine.

Popernack and Unger weren't ready yet to tie in. They wanted to wait until the water was closer.

Hall wanted no part of it at all, disdaining the thought of listening to his friends choking and drowning. He figured that when the water was about to cover him, he'd simply take a last breath, dive in, and surface where the water was at the roof, where he had nothing left to breathe.

The humor was black.

"How are we going to die?" one of the miners asked Unger.

"I don't know," he replied. "I could hit you on the head with a rock or else you could just drown."

Their spirit, bodies, minds exhausted, they settled in on an 18-by-30-foot damp patch of earth to await their fate.

An hour passed. They were still alive. Where was the water?

It seemed to have stopped its fatal advance. To confirm their hopes, Fogle put a stick in the mud to mark the water's edge and lashed a light bulb to it to act as a bobber in case the stick was inundated. The water didn't rise.

No longer resigned to the inevitability of death, they followed Fogle's orders and switched to rescue mode. Maybe they could hang on until help arrived. They continued to pound on the roof. They turned off their helmet lamps to conserve power. Every 10 minutes one or two of them would switch on their lamps to check the water level.

They were also fighting the cold. They covered themselves with canvas but it didn't help a lot. They were drenched and the mine temperature was only in the 50s. They sat back-to-back for warmth or lay on the ground, sandwiching workmates who would sometimes shake violently with chills. When spirits dipped, others would pick them up, often with humor.

The one who was consistently upbeat was crew chief Fogle. He took to heart that he was their leader, that it was his responsibility to bring up every man he led into a mine.

When one of them would say, "We're going to die," he was defiant.

"No!" he said. "Don't even let it enter your heads! We're getting out of here somehow, someway!"

Such resolve steeled the crew. So reliant were the eight miners on Fogle's leadership that they became dispirited when halfway through the ordeal he began coughing, throwing up, and complaining of chest pains. He said it was only the oily fumes from the compressed air aggravating his heartburn. They weren't sure.

Despite his physical ailments, he never flagged in his belief that they would survive.

That's why no one initially believed him when he returned from the water's edge to report it was receding.

"Don't be bullshitting us," they all said.

"I think it's going down," Fogle insisted. And it was.

For long stretches the men lay quietly in the darkness. Few slept for long. When they talked, they discussed anything they could think of, from what they'd do when they got out—attend a family reunion, one said—to their favorite foods, which were porterhouse steaks, T-bones, and ribs. Popernack asked his shivering crewmates what they would choose if they could: snuff, beer, or a hot chocolate.

Hot chocolate won.

Pugh, a snuff user for 35 years, had shared his tin of Timberwolf with his coworkers after theirs ran out. He tried to stave off the nicotine craving by thinking of things that made him happy, like the memory of bagging an 18-pound turkey on May 18. He thought that, even if he died, at least he had gotten that bird.

They had been staring in the dark for so long that Pugh started to believe that he could see his feet in the pitch blackness. And then Hileman said he could see his feet, too. The others told them they were crazy.

"I can even see a sky with stars, and a little town with houses and trees," Pugh responded.

At one point Hall's lunch pail was discovered floating about 100 feet from them. Inside they found the corned beef sandwich his wife had made him, still dry, and a bottle of Pepsi, which would supplement the 12 gallons of distilled water they had salvaged, normally used for machine batteries.

Every man took a bite of the sandwich except for Mayhugh, who figured the bite wasn't enough to end his craving for food, and Unger, who was afraid he'd vomit it up from the tension.

At another point Foy went scavenging and found two Mountain Dews on one of their machines.

They could hear the drilling getting nearer, brightening their hopes. And then at 1:50 A.M. Friday it stopped. Silence.

"Dear God, they gave up on us," Hall thought. "Dear God, please don't let them think we're dead and give up on us."

The others feared the same—except Fogle, their optimistic leader.

"Ah, they might have plugged up. Ah, they might have broken a bit," he said.

Fogle reassured the others that drilling would surely begin again.

CHAPTER SIX

THE RESCUE CONTINUES

As the miners grimly held onto their hope and their energy deep under the earth, the rescuers above them started Thursday evening imbued with optimism.

Schweiker predicted that the rescue mission was heading for the homestretch. He said the miners could be reached in half a day.

The super drill had arrived in pieces under a police escort with blaring sirens. Its assembly was under way, and the drill would be chewing through the earth by 7:30 P.M. Thursday.

Schweiker said a 240-foot life-saving tunnel to the miners could be dug by 3:00 or 4:00 A.M. Friday. Then it might take another three or four hours to remove the drill and lower a rescue basket through the 30-inch-wide hole.

In private briefings, Schweiker's advisers told him such a timetable was possible, but the super drill would have to slow down during the final 40 or 50 feet. Such a precaution would be necessary to prevent the life-saving air pocket in the mine from being destroyed. If the big drill bit punched through the roof too quickly, the air in the mine shaft would rush upward, potentially allowing water to fill the void and overwhelm the miners.

Even if the water didn't overtake the miners, the pressure release could cause another serious problem.

With the air being pumped into the mine, the miners were like scuba divers breathing pressurized air, and that could put them at risk for an illness called "the bends" if the pressure were to drop suddenly. In an extremely rapid pressure drop, in fact, bubbles

A SUPER DRILL: Workers keep a close eye on a super drill after its broken bit was fished out of the ground and a new one was attached. The drill eventually finished the job, breaking through the mine's ceiling at 10:15 Saturday night. Shortly afterward, voices from below were heard.

PETER DIANA/POST-GAZETTE

JOHN BEALE/POST-GAZETTE

A JOB WELL DONE: John Hamilton, foreman of the crew that drilled the shaft through which the miners would emerge, said he never gave up hope in the hours it took to bore through 240 feet of earth and rock. Hamilton, who was at the controls when the drill broke through, felt his stomach knot but steeled himself to remain focused on rescue efforts until the last of the nine miners was hauled up through what he called "the miracle hole." Then he cried.

could even form in their bloodstream and, like embolisms, kill them.

Dr. Nick Colovos, a reservist and physician with the Special Medical Response Team and Allegheny General Hospital, got in touch with experts at the U.S Navy in Norfolk, Virginia. On Thursday evening 10 portable hyperbaric chambers arrived at the drilling site to allow the miners to be immediately treated for decompression illness once they were brought up.

As another means of preventing a pressure drop, the response team's Dr. Richard Kunkle worked with engineers to create an instant airlock to place on top of the escape tunnel hole.

Larry Neff, construction supervisor at BethEnergy Mines Inc. in Revloc in Cambria County, pulled out a memo pad and roughed out two plans for a crude airlock, then phoned Don and Buddy Walker, a father and son team who operate Lincoln Contracting Company in nearby Boswell.

The Walkers fashioned a 40-foot-long, 3-foot-diameter tube with a sliding bottom door to go on top of the

Previous page

THE DAMAGED BIT: Rescue crews retrieve the 1,500-pound super drill bit that broke while tunneling toward the miners. Eighteen hours were lost before digging resumed on the miners' escape chamber.

ANNIE O'NEILL/POST-GAZETTE

ANNIE O'NEILL/POST-GAZETTE

DESPERATE HOURS: Volunteer firefighters and Somerset County residents wait for the rescue mission to regain momentum. The mood was glum with the rescue on hold again because the drill bit had broken.

escape shaft and keep the pressurized air from leaking as the escape capsule was being lowered to the miners.

The work on the airlock, as well as the sounds of drills, pumps, and people shouting, created a cacophony above the mine.

No tapping from the miners had been detected since 11:30 A.M. Thursday, but it easily could have been masked by the noise on the surface.

By now Schweiker was focused on the super drill and breaking through to the miners. He had a knack for speech that could be both casual and stirring. The super drill inspired him to use it. Before the sun comes up, he said, "We should be good to go. We're going to get our guys out of there."

Despite the governor's optimism, the mood at the fire hall was grim. Everyone's nerves were frayed..

The fire hall had become a one-stop shop for families and rescuers alike. Ministers provided counseling and held prayer circles; doctors passed out prescriptions for tranquilizers; the ladies' auxiliary brewed hundreds of gallons of coffee.

At one point the toilets in the fire hall broke down from overuse and portable toilets had to be brought in. The hall also took a beating near the kitchen, where the floor began to sag because of all the people coming through.

One of the people in the kitchen was Arlene "Sue"

Kovach, a feisty and energetic member of the fire department's ladies' auxiliary, who was consumed with the task of feeding about 600 rescuers and family members. Donations poured in, including pizza, sandwiches, ice cream, cakes, rigatoni with meatballs, and even 50 ham dinners with potatoes and green beans.

"Every day, we hoped the meal we were making would be the last," Kovach said.

But there would be more agonizing hours to get through before a final meal could be served.

About 1:50 A.M. Friday, after millions across the country who were following the drama had gone to sleep, the bit on the super drill broke. It was 105 feet into the ground.

Work on the escape hatch stopped cold. The broken bit had to be fished out before drilling could resume, a daunting job in itself.

Yost, of the company running the super drill, would later say that removing a broken bit can sometimes take a week. Schweiker knew it, too, but the thought was so depressing that he never voiced it.

With reporters cloistered two miles from the drill site, Schweiker said nothing publicly about the setback for five hours. Even so, the scent of disappointment was palpable.

KDKA-TV reporter Bob Allen went on air about 6:30 A.M. Friday and said he had heard of some snag or difficulty on the drill site.

About half an hour later Schweiker appeared live nationally on *The Today Show* and announced that the bit had broken and the rescue, for the moment, had ceased.

There were many low moments for families huddled at the fire hall, but this was the worst.

Schweiker made certain they knew of the setback before the public did, but that barely softened the blow. The fire hall had not yet been equipped with cots, so relatives of the miners spent the night outside, trying to sleep in their cars. What rest they got was interrupted when they were told of the calamity at the drill site.

Schweiker admitted it would be no easy job to retrieve the bit, but rescuers would try. Meantime, crews would start drilling a second rescue chute about 75 feet from the first.

V.W.H. CAMPBELL JR./POST-GAZETTE

HOPEFUL HUG: A relative of one of the miners embraces her son as they watch rescue workers race against time.

He prepared for his Friday afternoon visit to the fire hall by seeking an inspiring biblical passage to share with the families.

Upbeat as he walked in, Schweiker announced that he wanted everybody to pray with him. Then he read from Psalm 46: "God is our refuge and strength, always ready to help in times of trouble."

In later conversations with the miners' loved ones,

BILL WADE/POST-GAZETTE

A BRIEF RESPITE: A rescue worker rests on a piece of heavy equipment while other workers continue to drill downward toward the trapped miners. Lying on the ground is the yellow metal capsule that would be used to haul the miners up through the shaft.

Schweiker assured them that workers were doing everything mechanically, technically, intellectually, and humanly possible to revive the rescue.

Mary Unger, 87, whose son John was among the trapped miners, felt too frail to travel to the fire hall. From her home, she said what many of the miners' sequestered relatives were thinking.

"It's awful, the waiting. It seems like things just keep going wrong."

The man assigned to take news to the families was PBS leasing agent John Weir, who was serving as Black Wolf's spokesman. By Thursday night he was under orders from Schweiker to report hourly to the Sipesville fire hall, after making the rounds from the pumping stations to the mine to the drilling site.

"Basically, I told him it couldn't be done. I told the governor I needed an hour and a half. But he looked me straight in the eyes and said, 'It will be hour for hour.'"

"And I did that. Sometimes it was bad news because the bit broke, the drill broke, the water didn't go down."

It mattered not. To the folks at the Sipesville fire hall, Weir became their lifeline. No matter what they heard, either by cell phone or through the media, the families refused to believe any information unless it came from Weir, who was being fueled by adrenaline and Krispy Kreme donuts.

"The people down there at the fire hall, the numbers meant everything to them. A quarter of inch, an inch, two inches. Sure they wanted to hear big numbers. They wanted to hear the water dropped 10 feet. The drill went 25 feet. But that's not the case. It was inches.

"Every time I'd go to that fire hall, the number of eyes that were on me was unreal. It was like looking at the saddest children in the world waiting for me to tell them something. It was emotionally the hardest thing I'd ever done in my life. It was tough to tell. I had one of the mothers come up and put her arms around me and say to me 'I'm gonna hold you until you tell us the news.' "

Some of the most tense moments came when the first drill broke.

As the Yost people worked to fish it out of the ground and get the rig started again, a second drill was erected by Falcon Drilling. Operator Larry Winckler began punching through the earth. But his drill, too, would eventually break down.

Weir learned that news on one of his rounds. Winckler embraced him, crying.

"He said, 'John, I failed you. My rig's down and I really feel I failed you and I failed the people.' I said, 'Larry, you were my success story because when drill No. 1 wasn't drilling you gave me what I needed to tell the people. You gave me those inches and feet. And to them people down there, it didn't matter if it was No. 1 hole or No. 2 hole. You were drilling when No. 1 was down and that's what the people wanted to hear.' "

While the second drill was pounding through the earth, drilling crews were struggling to extract the broken bit in shaft No. 1. They needed a specially made "fishing tool" to do the trick.

Rescue crews heard that Frank Stockdale, plant manager of Star Iron Works in Big Run, Jefferson County, could build just what was needed.

Stockdale was happy to try, but first he needed engineering prints of the drill. They would give his workers a road map of how to attack the bit.

Normally, a job so big would take three or four days. Rescuers faxed the prints and implored Stockdale to pull out every stop to get the work done sooner. His 95-member shop did just that, building the tool in three hours.

Normally, a job so big would take three or four days. Rescuers faxed the prints and implored Stockdale to pull out every stop to get the work done sooner. His 95-member shop did just that, **building the tool in three hours.**

"When they call you up and tell you a National Guard helicopter will be waiting to pick it up when you're done, you get a sense of urgency," Stockdale said.

He was sure the tool would work, and he was right. This specially made fishing hook grabbed hold of the 1,500-pound bit and yanked it from the hole at about 4:00 P.M. Friday.

After a 14-hour shutdown the first rescue tunnel was back in business. But drilling on the first escape hatch would not resume until about 8:00 P.M. Workers kept digging the second one, too.

The pace was nowhere close to the one that Schweiker had predicted a day earlier. But hope, once as fragile as bone china, had been restored to the families. The fire hall resounded with cheering.

Once things settled down, a soberness took hold. By now, everybody knew that reaching the miners was not something a governor or anybody else could predict.

CHAPTER SEVEN

THE MEDIA CONVERGES

By the time the super drill accident had occurred, media from around the nation had arrived in Somerset County to chronicle what had become an international drama.

But the reporting didn't start out that way.

It really began with a reporter at the *Somerset Daily American*, 23-year-old Leona Kozuch. On her way out the door Wednesday, sometime after 10:00 P.M., Kozuch asked the copy desk if there was anything else to do.

"They passed me an e-mail from County Control, which is our 911," she said. "It was real brief: 'Nine miners injured in a mine near Quecreek.' I live right near there, so I said I'd stop by and see what was going on."

Two Pennsylvania State troopers guarded the entrance to the mine works as Kozuch pulled off the two-lane highway near her home.

There were no other cars nearby. Behind the policemen, the mine entrance was dark, with no apparent activity. The officers told her to wait for a press briefing at the nearby Christ Casebeer Lutheran Church, located directly over part of the Quecreek Mine.

No one else was there when she arrived.

"I waited for a while and then ran down to my parents' house," she said. "My mom told me that my uncle worked in that mine and that there had been an accident."

Kozuch called her uncle for details, phoned the news to her paper, and returned to the church. For the second time in less than a year a Somerset County newspaper reporter was first on the scene at one of the hottest stories in the world.

TV crews from Johnstown and Pittsburgh joined Kozuch and a few other print reporters for the 1:00 A.M. press conference at the church. The story was still a local phenomenon.

But by Thursday afternoon, the drama unfolding above and below ground had attracted the national

BILL WADE/POST-GAZETTE

QUESTIONS AND ANSWERS: Pool reporters and onlookers, including Bill and Lori Arnold, foreground, surround David Hess, secretary of the Pennsylvania Department of Environmental Protection, just hours after the successful rescue was completed. The rescue site was located on a farm owned by the Arnolds. In the background are the drilling rigs, including one topped by an American Flag.

press corps, and by Friday, TV satellite trucks were beaming the news to viewers as far away as Europe and Australia.

By the weekend, nearly 200 media members were on the site, including crews from CNN, MSNBC, and Fox News Channel and from such newspapers as the *New York Times*, *Chicago Tribune*, *Los Angeles Times*, and *Atlanta Journal-Constitution*.

Competition to break news was fierce. Police threatened to jail reporters who pushed too hard for access to the miners' families, sequestered in the Sipesville Volunteer fire hall.

Local print reporters sometimes felt cut out of the loop. After being bumped off a press corps bus heading to the rescue site to make room for an NBC crew, Kozuch complained so much "they brought the bus back just for me and my photographer."

"That's just a part of covering a big story," said the Johnstown *Tribune-Democrat*'s Mike Faher. "The same thing happened after Flight 93. It's frustrating

The media gathers. AP PHOTO/DAVID SWANSON/PHILADELPHIA INQUIRER

when you're trying to have a conversation and a TV crew interrupts for a sound bite."

TV reporters struggled to get the pronunciations right (locals call the nearby stream the KWEE-mahoning Creek, but call the work site the KEW-creek Mine).

The reporter with the most familiar face, Fox News' Geraldo Rivera, didn't arrive until 8:00 P.M. Friday. When he got there he expected to be covering a tragedy.

"I got there at sort of the nadir, when they hadn't heard from the miners in a while. My first interview was with a local woman who was beginning to fray around the edges. She was harshly critical of [Governor Mark Schweiker] and insisted on speaking to me. She told me she was the cousin of one of the miners and the family was getting very edgy and impatient."

Rivera, too, was growing impatient with the governor's guarded announcements to the press.

"The governor at that time seemed more of a cheerleader than a soothsayer," he said. "It was difficult to cover—instant history. You want to be the first to get it, but the consequences of not getting it right are much greater than the consequences of not getting it first."

The worldwide attention brought an instant response. Dozens of Web surfers discovered the Sipesville Fire Department's e-mail address and sent messages of support. After Larry Baughman, manager of the Summit Diner, was interviewed on Fox, he received calls from people around the world, most of whom asked to pray with him. Among them was a woman from Alabama, where 12 miners were killed last fall trying to rescue a fellow miner, who also died.

With the exception of the rescue workers frantically drilling an escape route, that's all anyone could do at that point: pray, and wait.

AP PHOTO/STEVE HELBER, POOL

SOUNDS OF LIFE: It was the moment everyone had been waiting for. A rescue worker drops a line to establish communication to determine if the Quecreek Nine were still alive.

CHAPTER EIGHT

BREAKTHROUGH!

The breakthrough came Saturday at 10:15 P.M.

After days of effort and a broken bit, the rescue drill punched through into the trapped miners' dank quarters. The drill rig operator pumped his fist in the air, then jumped up and started yelling. The escape shaft, through which the capsule carrying the miners to safety would travel, was finally in place.

The moment of breakthrough—the instant people above ground had waited for like 1969 America waited for the astronauts' first steps on the moon—wasn't initially noticed by the miners, 25 stories underground.

Once steady pumping had caused the water around them to recede, the miners had begun taking turns walking 250 feet down the passageway every 10 or 15 minutes to pound nine times on the six-inch air pipe and check the area where the drilling sounds were coming from.

Saturday at 10:15 P.M. Hileman and Foy made the trek. Their cap lamps were dim and just about out of juice.

That's when they found the drill opening.

Back on high ground the other miners were lying down, trying to stave off the cold, when Hileman came bounding back.

"We found the hole!" he screamed. "Everyone get down there!"

No one needed a second invitation. They bolted toward Entry No. 4 with energy they never knew they had.

Hileman then found Unger, who was separated from the rest of the group. "You want to go home tonight, John?" he asked casually.

"Yes, I wouldn't mind going," Unger replied.

"Well, grab your stuff," Hileman yelled. "We found the hole!"

Mayhugh unbuckled his mining belt as he ran toward the hole. He knew he'd never use it again.

The drill had touched down about 300 feet away, across two crosscuts.

Officials asked the families whether it would be okay to broadcast the rescues live. Foy was the second person to raise her hand indicating her agreement. She thought people around the world had a right to see **what their prayers had done.**

V.W.H. CAMPBELL JR./POST-GAZETTE

HOPEFUL: Megan Mento, nine, says a prayer for the trapped miners. She was visiting Lori Arnold, whose farm is over the mine.

When the miners got there they began yelling, "Get us out! Help us! Please get us out."

Back above ground, rescue workers took a pressure reading at the top of the big hole and got a zero, indicating that the pressure below was normal and they wouldn't need to use the customized airlock.

Then they heard tapping.

The rescue workers, knowing a pool of reporters and photographers had been allowed to enter the drill site, tried not to show any emotion. They didn't want to violate Schweiker's edict that the miners' families should learn of any developments first. By radio, a worker tersely asked someone from the command center to join them—fast.

Yet they couldn't contain their joy. They began high-fiving each other and smiling.

After getting the radio message, Stricklin, the federal mine safety official, ran out to the site thinking someone was hurt. When he heard the miners were alive, his heart pounded.

Nearby, Rob Zaremski, who had been on site for more than a day, had been waiting to use a special pen-shaped, two-way communication probe—the CON-SPACE Rescue Probe—carried by his company, Targeting Customer Safety.

He was thrilled when a DEP official told him to bring his equipment to the site.

The probe, containing a speaker and microphone housed in a stainless steel compartment, would send sounds to Zaremski's headset.

Wearing a yellow hard hat, Zaremski slowly lowered the probe into the six-inch air pipe. He had attached a child's glow stick to it so it would be visible in the dark mine.

After the probe had descended 75 feet, he began saying, "Stay where you are. Can you hear me?"

At 100 or 125 feet, Zaremski thought he heard people saying, "I can hear you." But he didn't know whether it was people topside thinking he was doing a sound check.

Then, unmistakably, he heard, "We can hear you."

Phillippi was at the other end of the device, speaking into the slender microphone.

To Zaremski, hearing the voice was like saying a prayer and hearing God answer. Surprised, he asked three more times if the person could hear him, just to be sure.

Ray McKinney, the federal mine safety administrator, fed Zaremski the next questions.

"Are you the trapped miners?"

"Yes, we are."

"Are you OK?"

"We're OK except for our boss is having chest pains."

That's when Zaremski gave the thumbs up captured on TV and sent around the world.

"How many are you?"

"We're all nine here," came the reply.

Zaremski raised nine fingers and grinned. As word spread up the hill from the rescue site, small spontaneous celebrations broke out.

Word of the breakthrough quickly spread through a media-briefing area five miles from Somerset, even before the governor had confirmed it.

Eager to share the news with the miners' families, Black Wolf Coal Company executives John Weir and Dave Rebuck jumped into a waiting state police car.

"Give them the road," an officer said as they sped away. "Get them there."

The trooper behind the wheel complied.

"I haven't gone that fast since I was 16 years old and I had a good running Chevelle," Weir said. "I have to say, that trooper could drive."

As they pulled up to the fire hall with the sirens blaring, people came running from all directions.

Weir made his way to the stage, but before he delivered the good news he asked those assembled to promise that no one would rush to the drill site.

"They all gave me the nod. It was like 200 people. I said, 'There's nine alive.'"

The hall exploded in cheers, applause, and shouts of "Praise the Lord!" that didn't stop for 10 minutes.

Miner John Unger's son, Stephen Unger of Atlanta, had never seen people so emotional, and the 25-year-old had never cried so hard in his life.

Denise Foy, the wife of Tom Foy, tried not to show her emotions. Every time she had cried before, her four daughters had cried, too, especially Leslie, wife of Blaine Mayhugh.

When the good news came that nine were alive, she just grabbed her daughter and squeezed.

Officials asked the families whether it would be OK to broadcast the rescues live. Foy was the second person to raise her hand indicating her agreement. She thought people around the world had a right to see what their prayers had done.

Back at the site, Phillippi asked Zaremski, "What took you guys so long?"

McKinney grabbed the headset from Zaremski.

"Who am I speaking to?" McKinney asked.

"This is John," Phillippi replied.

McKinney was so happy, he wouldn't have minded if Phillippi had said his name was Greta.

"Are all nine at the same location?"

They were.

Phillippi asked for flashlights, food, cap lamps, snuff, and chewing tobacco. He also told McKinney the drill had deposited a pile of dirt that would have to be moved away from the opening before the rescue capsule could be dropped through.

After inquiring about their health, to make sure no one's life was in danger, McKinney told the miner the rescue capsule would be coming down the shaft.

Phillippi told McKinney the largest person in the mine was almost six feet and 230 pounds.

Earlier, McKinney had spotted a large rescue worker—6' 3" and 310 pounds. McKinney had asked him to climb into and out of the rescue capsule. He was able to do it. If he fit, McKinney thought, surely the largest of the nine miners would too.

Now McKinney handed the headset to Dr. Richard Kunkle of the medical response team. He asked about injuries and took medical histories of the men.

Phillippi again asked for snuff and chewing tobacco. Kunkle agreed as long as Fogle, who had been suffering chest pains, didn't use any of the tobacco.

At another point mine owner Rebuck telephoned down his own message to miner Hall.

"Hey Denny, you're a hero," he said. "Your phone call saved those nine other guys."

Hall, who for 78 hours hadn't known the fate of the other crew, cried when he heard the news.

Finally it was time to send down the steel mesh rescue capsule, an 8½-foot high cylinder with lantern lights attached at the bottom.

On its first trip it carried flashlights, cap lamps, Hershey and Kit Kat candy bars, blankets, raincoats, drinking water, glow lights, Skoal and Copenhagen snuff, and Mail Pouch tobacco.

After it got there Pugh packed so much snuff between his cheek and gum and wolfed down so many candy bars that he felt dizzy.

Up top, another miner at the site joked with the trapped nine, telling them their families had been told they were alive and "most of them are happy about it."

Then federal mining official Jeffery Kravitz took the headset to decide the order in which the miners would be raised to the surface.

They were told to put Fogle on first, something the miners already had agreed to among themselves. After that they would come up in order from the heaviest to the lightest. The last person had to be the smallest because he wouldn't have anyone to help him get in the capsule.

As the rescue operation got under way, the news of the miners' miraculous survival spread far and wide.

At Buffalo Blues restaurant in the Pittsburgh neighborhood of Shadyside, the bar erupted in a huge cheer when people realized the miners were safe.

When members of the Somerset High School class of 1967, who were having their reunion at the Oakhurst Tea Room, learned of the imminent rescue, the 90 class members abandoned the ballroom and the DJ and headed to the bar to watch the rescue on television.

At the fire hall, e-mail began to arrive.

A woman whose firefighter nephew was killed at the World Trade Center wrote, "Please relay to [the miners] and their families that they were never alone. All of Brooklyn, New York, was with them in prayer . . . Now they are prayers of thanks."

Even the press area erupted with joy as journalists forgot they were supposed to be jaded and coolly neutral.

The first reporter on the scene, the *Somerset Daily American*'s Kozuch, cried when she heard the news.

Pamela Mayer, the publisher of the Johnstown *Tribune-Democrat*, had never before worked on a breaking news story with a happy ending. She wasn't even going to complain about the overtime.

And Fox News' Geraldo Rivera out Geraldo-ed himself, gushing as he delivered the news. He hugged the men he interviewed and kissed the women. He even called Schweiker "Dude."

CHAPTER NINE

UP, SAFE AND SOUND

The medical workers breathed sighs of relief as some of their biggest concerns vanished.

The water level in the mine had dropped, simultaneously reducing the air pressure, so the untested airlock that had been invented and built on the spot would not be needed. Instead it would be donated to federal mining officials for study and possible future use.

The miners were alert and able-bodied, so they could climb into the 22-inch diameter rescue capsule without help.

Through the communications person on the surface, medical team leader Danny Sacco gave instructions as each miner rose through the 240-foot drill shaft. What's your name? What's the worst thing wrong with you? Don't try to get out by yourself.

Sacco, standing at the bore hole, greeted them with a "welcome home" as he and another rescuer helped them out of the capsule. Some replied, "It's good to be home."

Fogle, who had complained of chest pain, would be the first to surface, at 12:50 A.M. on Sunday. As with those to follow, the whites of his eyes shone from a face black with coal dust.

They looked like drowned rats, Sacco said. But that didn't mean the miners had crouched in water during their entire ordeal. Instead, the drill shaft had gone through an aquifer, drenching the miners in yet another torrent of cold water in their final exits. After discovering what had happened to Fogle, raincoats were sent down the capsule to shield the others.

The other eight miners were brought up according to weight, heaviest to lightest.

Mayhugh was second. The others helped him squeeze into the tight-fitting capsule. Groundwater poured down so heavily that he had trouble breathing. He couldn't hold the microphone nor keep the headphones on. He could hardly hear the rescuers. He'd had enough.

"Get me the hell up and I mean now! Let's go!" he shouted.

ON TOP OF THE WORLD: Rescue workers cheer and applaud when Blaine Mayhugh is pulled to safety. Just 15 minutes later Mayhugh's father-in-law Tom Foy, also would be rescued. Mayhugh landed a guest spot on David Letterman, and there he drew wild cheers and applause once again.

MARTY GINTER,
COMMONWEALTH MEDIA SERVICES

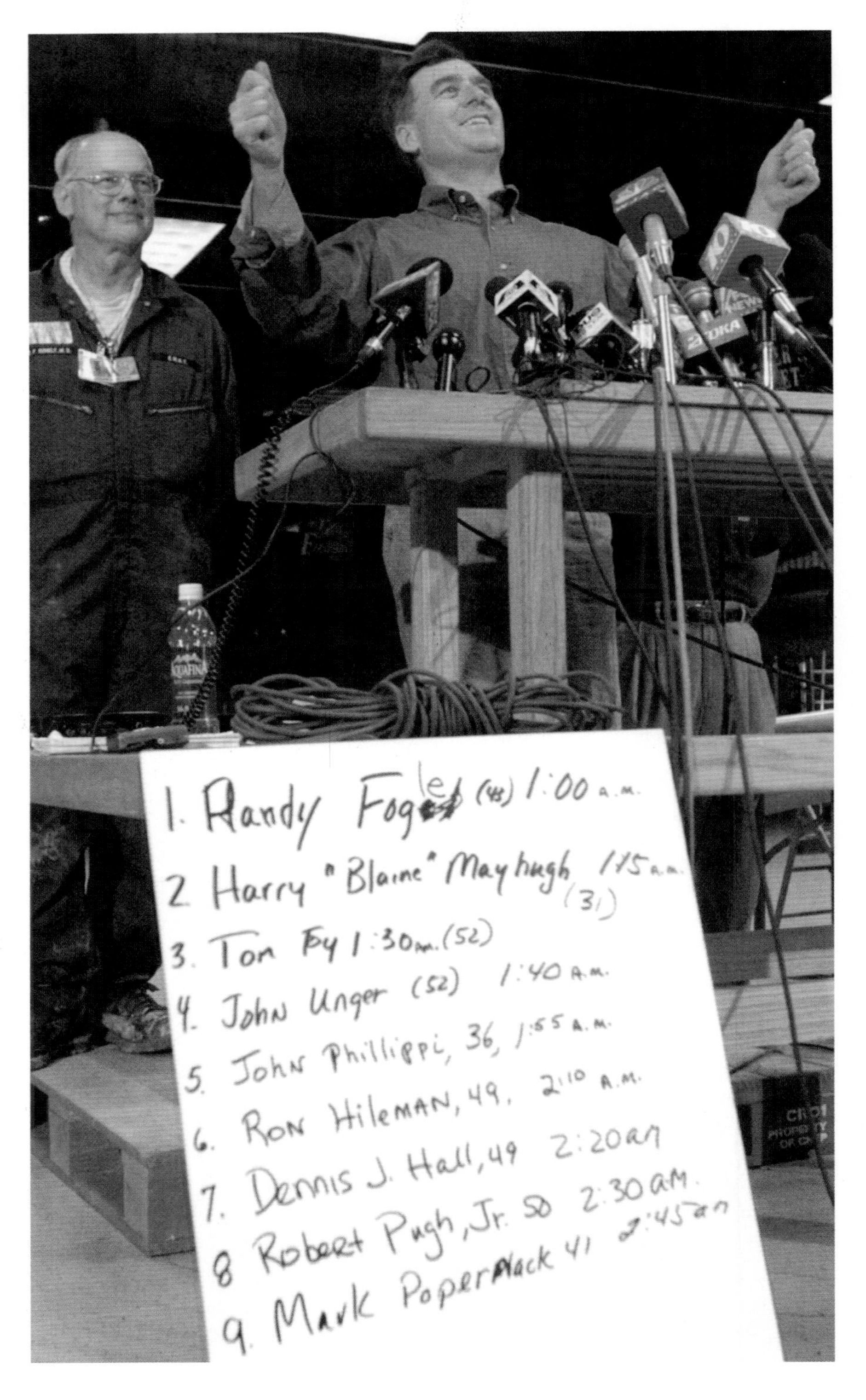

JOHN BEALE/POST-GAZETTE

JUBILATION: An animated Governor Mark Schweiker talks to the press at 3:30 A.M., just 45 minutes after the last miner was rescued. Schweiker drew on his years of experience as head of the Pennsylvania Emergency Management Agency to become a force behind the rescue and the face at the frontlines. "There was no red tape when he was around," said Black Wolf owner Dave Rebuck. "Without him it would not have gotten done." Behind Schweiker is Dr. Richard Kunkle, of the Special Medical Response Team, who evaluated each man as he was brought up. In the foreground is the poster board where each man's name, age, and time of rescue was listed.

AP PHOTO/STEVE HELBER/POOL

FIRST ONE OUT: Rescuers help Randy Fogle exit the capsule that lifted him to safety. Fogle, who was suffering from chest pains, was brought up first because of his condition. The rest were brought up according to weight, heaviest to lightest.

Once the capsule began its ascent, Mayhugh calmed down. About three feet from the surface, he could see the bright lights and hear the applause and cheering.

His heart raced and he began to cry. On the surface, everything was a blur until he was placed on a stretcher. Schweiker leaned over him. "I just talked to your dad 10 minutes ago," the smiling governor said. "He told me, 'Get my son out of there.'"

Mayhugh's father-in-law, Foy, was the next one out. Unger, Phillippi, Hileman, and Hall followed.

Then came Pugh, already a little dizzy from the Hershey bars and snuff.

But even more disorienting were all the lights and cheering as he, second to last, rose to the surface. "It just gave me a feeling like the Steelers won the Super Bowl."

Popernack, at 2:45 A.M., was the last man out.

Sacco's counterparts elsewhere later told him that

AP PHOTO/GENE J. PUSKAR/POOL

LAST MAN OUT: A medic team shuttles Mark Popernack to a treatment station, where his medical condition assessed. He was the last of the nine miners to be lifted from the mine.

the transfer of miners from the capsule to the treatment units was so smooth that it looked rehearsed. It was. Sipesville firefighters and medical team members had trained for the transfers before the rescue.

The miners could not be jostled on the journey from the drill site up the hill to the decontamination unit and beyond because bumps can trigger heart rhythm problems in abnormally cold patients. Physicians Kunkle and Dixon took turns escorting stretchers in case an emergency occurred.

Rescuers at the drill site brought up the miners in 15-minute intervals rather than the two hours each that had been predicted. Soon the medic team had to hightail it back down the hill to pick up the next miner after dropping one off at the treatment stations.

A Somerset company had organized the efficient decontamination procedure, in which clothes were cut off and special detergents were used to wash away the oily, flammable coal dust, even from the miners' ear canals. The residue would be dangerous in the potentially explosive environment of a hyperbaric chamber.

Each miner was assessed by a physician and a physician's assistant. A nurse started an intravenous line to deliver fluids. Vital signs, an EKG, and a reading of blood oxygen content were taken. A Navy corpsman did an exam solely to check each miner for the bends.

It turned out that none of the miners needed the pressurized chambers at the drill site. Instead the miners were transferred either by helicopter, flying at low altitude to diminish pressure problems, or by ambulance to hospitals.

Fogle, Foy, Unger, Phillippi, Hall, and Pugh were sent to Conemaugh Memorial Medical Center in Johnstown. Mayhugh, Hileman, and Popernack went to Somerset Hospital.

Dr. Russell Dumire, a trauma surgeon, waited on the helipad at Conemaugh to escort the first patient to the emergency room. He has performed surgery in a hyperbaric chamber and is a diver, and his grandfather was a mine engineer who performed rescues. Dumire happened to be the surgeon on call that weekend.

The medical staff was thrilled to find the miners were smiling, alert, and talking. They were also shivering. They had been wrapped in warm, dry covers after the decontamination, but by the time they got to the hospital, residual water on their bodies became almost ice cold and the blankets were saturated.

"They were freezing cold," Dumire said. Fingers, feet, and lower legs were purple and mottled from immersion. "It looked like if you rubbed real hard against their feet, you could rub the skin right off."

The lowest body temperature the medical staff found was about 92.5 degrees, the warmest 96.8. Normal is 98.6.

The hospital staff replaced cold, wet blankets with dry ones at their backs and bearhugger wraps in front to envelop them in heat. A second IV was started to push more warm fluid into their systems.

After that they were moved to another area of the emergency room to wait for transfer to a hospital ward. It was there that each miner was reunited with his family.

"They were huge groups," Dumire recalled. "Some had 12 to 15 family members."

Pugh was overcome in what he called the "drama room"—the trauma room at Conemaugh—when his girlfriend, Cindy Thomas, and relatives rushed in, as he put it, "like a herd of cattle."

"I only cried twice in my life: when my son was born, and that day there."

Denise Foy said that when she finally got to see her husband, "He just grabbed my hand and we started kissing. We didn't really say nothing."

The Foys' daughter, Leslie, got to Somerset Hospital 10 minutes after her husband, Blaine Mayhugh, arrived there. They embraced, making up for that kiss he'd neglected to give her days before.

Back at the drill site euphoria had set in. Mine owner Rebuck hugged the people at the rescue site. He hugged the people at the mine site.

"Once I knew we had nine guys, I sat down and watched. This is history and I'm going to enjoy this," said federal mining official Stricklin. "There wasn't a rescue worker on the site who didn't cry. I'm sitting there thinking, 'Life is good' and I'm about ready to start singing that 'Kumbaya' song. I wanted to bottle that moment and never have it end."

Joining him on a row of hay bales were fellow federal official Urosek and state mining expert Sbaffoni.

"It was just so great . . . all nine guys," Urosek said. "Joe Gallo [a PBS engineer] had done a lion's share of the work and I thought he was going to pass out."

"We shook, we hugged," Sbaffoni added. "Jesus, I'll tell you, it was something. You're totally exhausted, and then your emotions . . . well, you're done. I still can't stop crying."

Urosek and Stricklin finally stirred, gathered their equipment, and drove home to Fayette County. Sbaffoni joined the governor for a final news conference at 3:30 A.M., and was ready to leave the site at 5:00 A.M. He was too tired to drive.

"I'd seen this bed-and-breakfast before, and I went down to Somerset to see if they might have a room," Sbaffoni said.

"But it was so late, and I'm debating whether to knock. Then the door opens and the innkeeper is there. He said, 'I was watching TV and I waited up for you.' "

Sbaffoni slept for a few hours, then drove home, arriving at 1:30 P.M. Sunday. After a failed attempt to sleep he decided to join his softball team in Collinsburg for a scheduled game. As he walked onto the field, applause broke out. In the stands, he saw a sign: "Thanks, Joe!"

After attending a final press conference, Kunkle drove home to New Florence, where he slept for most of the day. Despite all the shut-eye, he realized that his head was still fuzzy when he sat at his computer and couldn't remember how to work his e-mail program.

At 12:55 P.M. Sunday, Schweiker walked into Foy's hospital room, but he caught the miner at an inopportune moment. **"Can we make this quick?"** Foy asked. **"My NASCAR race is going to start in five minutes."**

Sacco led the medical response team trucks back to their base in Indiana County, driving slowly and occasionally pulling over to rest. Park the trucks, drop everything, and go home, he ordered when they got back at 6:30 A.M. Cleanup could wait.

"We've waited for 20 years for the opportunity to bring live people out of a mine," he said. "This truly is a lifetime experience."

All the miners except Unger, Fogle, and Foy went home from the hospital on Sunday.

Doctors worried that Unger, who complained of right-shoulder pain, was suffering from the bends and not the arthritic pain that had plagued him in the past.

"I had some problems with it before, but it seemed that when we came out of there [the mine], it just throbbed," Unger said. "The pain was astronomical."

After consultation with the Navy experts, and not willing to take chances, the medical staff opted to put Unger in one of the portable hyperbaric chambers, which had been brought to the hospital for just such an instance, on Sunday afternoon.

Unger "dove," as the doctors put it, for six hours. The pressure was initially set high to dissolve the nitrogen bubbles in the tissue and then gradually reduced to allow the gas to dissipate harmlessly.

"That was horrible; I didn't like that," Unger said of the hot chamber. But "it made my arm feel great, so I guess it did the job."

Fogle and Foy, both of whom had chest pain at times while trapped in the mine, had to stay for further cardiac monitoring. Fogle had an irregular, rapid heart rate that was treated with medications. And the fumes from the air pump had aggravated longstanding heartburn and throat problems, triggering fits of coughing and throwing up.

Fogle said his heart is fine now, but he still can't eat. He went home on Monday, as did Unger.

Foy has a history of heart problems and once had an angioplasty for clogged heart arteries. He was pain-free during his hospital stay, but the results of a stress test warranted further evaluation, Dumire said. After echocardiograms and a procedure to check for blocked arteries, he was discharged late Tuesday evening.

Because the world was anxious to hear how the miners had fared during their 78 hours underground, several press conferences were arranged at the hospital.

"I came today to thank everybody out there," said Unger at one of the gatherings, expressing a shared feeling among the men.

Even with all the attention, the miners tried to keep a sense of normalcy. At 12:55 P.M. Sunday, Schweiker walked into Foy's hospital room, but he caught the miner at an inopportune moment.

"Can we make this quick?" Foy asked. "My NASCAR race is going to start in five minutes."

CHAPTER TEN

THE AFTERMATH

As Sunday evening approached, heralding a return to normalcy in Somerset County, most of the rescued men said their mining days were over, and it was hard to find anyone who faulted them for that.

"My husband said he'll never go into a mine again unless something like this happens again and he has to rescue someone else," said Foy's wife.

Hall was one of the few who said he might continue in the profession. Others said they might not have a choice, because it's all they've ever done.

The MicroPower Institute of Technology in New Kensington offered to retrain the miners for free in computer or systems repair. That would allow them to take jobs paying in the low to mid-$20,000 range, much less than they had been making as miners.

More lucrative offers from movie studios and book publishers soon came along.

Interest flowed in from Paramount Pictures, Columbia Tristar/Sony Pictures, and the Walt Disney Company, with Disney winning the competition by wrapping up a $1.5 million TV movie deal in early August.

The company will produce a television movie for ABC that probably will air in the 2002—2003 season. Wilkinsburg lawyer Thomas Crawford helped the miners seal the deal, which will bring the miners and one of the rescue workers $150,000 each.

Andy Glasscock, a sergeant in the Midland, Texas, police department, had a sense of what the miners might be experiencing with all the attention from Hollywood. Glasscock was a central player in the dramatic 1987 rescue of 19-month-old Jessica McClure, who fell through an eight-inch-wide abandoned well opening.

Within a week after Jessica was rescued, producers came calling, Glasscock said, and the rescuers began to argue over how they were to be portrayed in a TV movie. Two months of bickering divided the community.

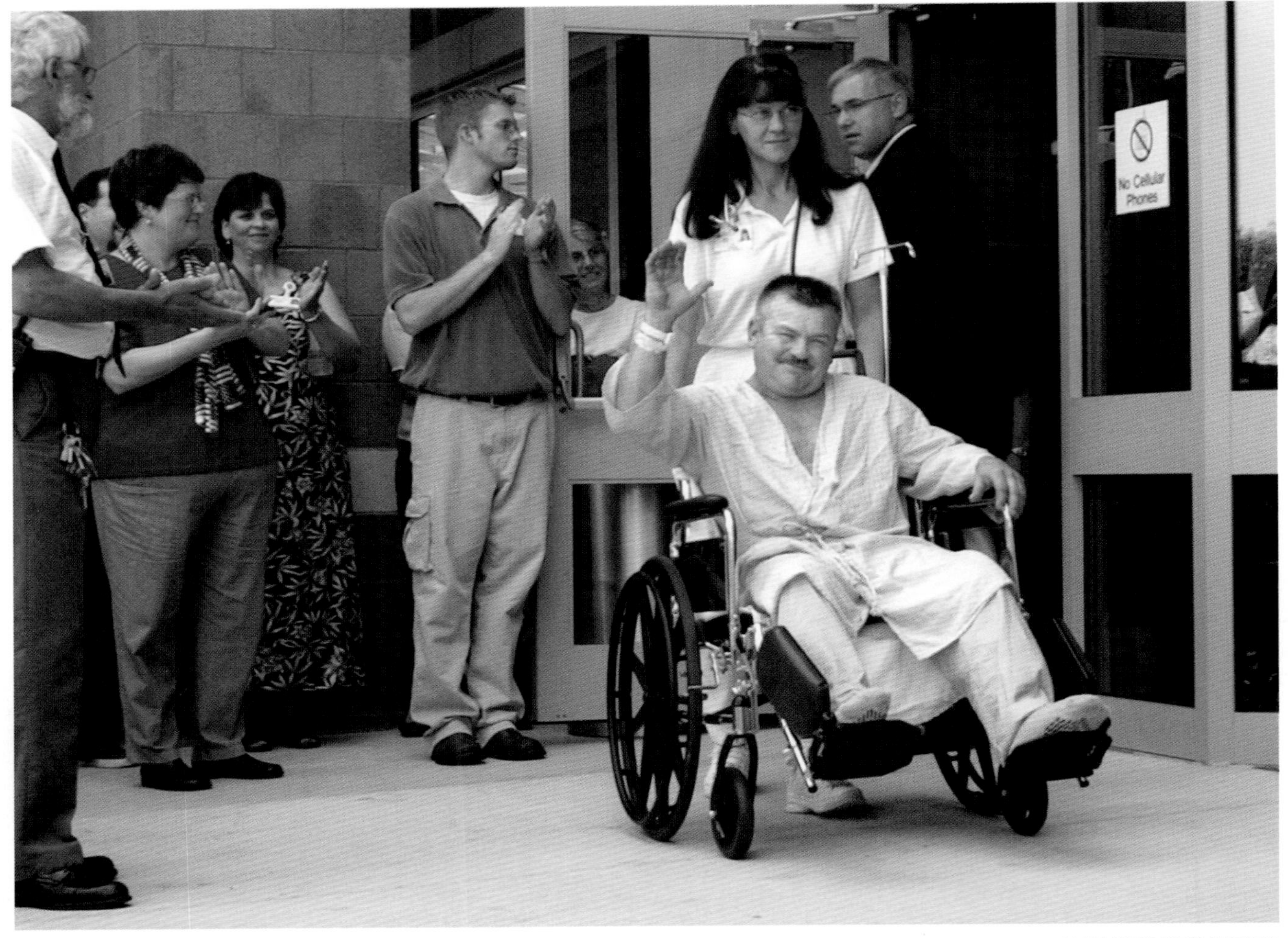

MATT FREED/POST-GAZETTE

BACK ON TOP: Tom Foy waves to the waiting media throng at Conemaugh Memorial Medical Center the day after the rescue. During a press conference, Foy said he doubted that he'd go back to his mining career of nearly 30 years, because "maybe it's just too much." His wife, Denise, chimed in: "He ain't going nowhere underground."

"It was a disaster," Glasscock said. He suggested that the miners savor their rescue and not worry about "damn Hollywood."

Pugh, who suffered from swollen feet and ankles after being on his feet in the mine for so many hours, said he hid for three days after emerging from the depths. Being pursued by the media made him feel like a "fugitive," he said, but things had quieted down by the end of the week.

Fogle, the crew boss, was still having trouble eating because of the heartburn that had been aggravated in the mine. He wasn't sleeping much, either. A number of times he went outside his house near Garrett to watch the sun come up and think about the "second chance" he had been given after being so near death.

As the Quecreek Nine tried to come to terms with their feelings after the rescue, miners outside the spotlight began descending into the very place that had threatened their colleagues. Eric "Snook" Brant of Somerset was one of six miners who went into the entrance on Sunday night to start the long process of de-watering and refurbishing the mine.

"I'd be lying if I said it wasn't a little bit scary," Brant said.

State and federal agencies had already launched investigations into why maps of the Saxman mine were inaccurate, but investigators said they would not be able to get inside the Quecreek mine for a closer look until water levels receded significantly.

Five days after the rescue, two Democratic state lawmakers asked Attorney General Mike Fisher to convene a grand jury to look at possible criminal misconduct by mine officials or state regulators who knew beforehand that there were unmapped mine voids near Quecreek. Fisher said the next day that it was too early to determine whether there was evidence of criminal wrongdoing, but he planned to inspect the mine.

The cleanup at Quecreek was expected to take several weeks.

Within a week of the rescue the process had already been delayed; state environmental officials halted the pumping of water from the mine after finding excessive amounts of iron flowing into nearby Quemahoning Creek. The pumping resumed several days later after Black Wolf Coal Company built treatment ponds.

MARTHA RIAL/POST-GAZETTE

GOOD TO BE HOME: Miner John Unger receives a hug from eight-year-old neighbor Cheyenne Alwine after returning home. Unger, fourth of the nine miners to be rescued, lives on a farm in Hollsopple in Somerset County.

The mine, which has enough coal to support another 15 years of mining, is expected to open

JOHN BEALE/POST-GAZETTE

NINE CANDLES: Servers Nick Morgan and Samantha Miller light one candle for each of the miners before Sunday Mass at All Saints Parish in Boswell, Pennsylvania, where miner Robert Pugh worships. The miners were rescued during the predawn hours that day. The altar was decorated with nine mining helmets, nine red carnations, and nine statues of miners, in addition to the candles

again, but no one is sure when. In the meantime all 63 employees—except for the 9 who were rescued—went back to work to help with the cleanup.

Still to be settled up is the bill for the rescue work, which may be around $10 million. At this point it's hard to say who will shoulder the cost, and how much of the effort will be donated by individuals and companies.

Hundreds of curious onlookers went to the Arnold farm in the week after the rescue, parking their cars by an abandoned fruit stand across the road and walking down the narrow lane adjacent to the rescue scene. From there they gazed at the mound of rocks covering the drill hole and admired the tiny American flag that had been placed on top.

Foy's younger brother, Wilbur, served as a volunteer guard at the site on Tuesday. He willingly spoke with onlookers who wanted to learn more about his brother. Pugh also came back to the site with his family that Tuesday.

Unger stopped at the rescue scene in his pickup truck that day after driving around "to see what I had

JOHN BEALE/POST-GAZETTE

JOY AND GRATITUDE: Pennsylvania Governor Mark Schweiker, who was occasionally near tears during the frustrating moments before the rescue, celebrates success with miners, from left: Robert Pugh, Blaine Mayhugh, Thomas Foy, and Ronald Hileman. They had gathered at the Green Tree fire hall to meet with President Bush.

missed" and stopping for an ice cream.

"I was just looking," he said, "and a lady there starts to tell me about these nine miners who were rescued there."

He listened for a moment and then confided to the woman, "Ma'am, I was the fourth of nine."

Keith Bowser and his wife, Betty, came down to the rescue site from Butler, Pennsylvania, on Wednesday. They asked Doug Custer, who escaped the mine early on and who was guarding the drill hole, if they could take his photo. He said sure.

"I was a normal person before this, and I want to be a normal person after," Custer said.

Bowser said there ought to be a memorial at the site, but as of last week, there was none. Mark Zambanini, chief of the Sipesville Volunteer Fire Department, suggested that people simply put up a "Nine for nine" marker and then move on.

"Everybody did his job and now it's over," he said.

But moving on could also mean a new fire hall for Zambanini's department.

Country music star Travis Tritt, who was moved by the rescue effort and stunned to hear about the poor condition of Sipesville's hall, gave the department $25,000 for its building fund. He also asked his fans to donate and promoted the cause on his website.

Before the rescue the firefighters had saved $25,000 for the renovation. The project is expected to cost $1.2 million.

A taped interview with all the miners appeared on NBC's *Dateline* on the Tuesday after the rescue. During the interview a few of the miners cried. Many shook their heads. When Fogle brought out the muddy bucket that contained the men's final notes to their families, the miners swore it would never be opened.

Also that night, Schweiker appeared on *The Tonight Show*. Some observers have likened Schweiker's actions during the crisis to those of former New York City mayor Rudolph Guiliani after September 11. But on the day of the rescue, Schweiker, sleep-deprived and still sporting jeans, brushed aside the praise, saying he had overseen emergency management as lieutenant governor in the administration of former Governor Tom Ridge.

"International attention or not, it's been my job for eight years," he said.

The Quecreek experience, which raised his political profile, hadn't altered his plans to get out of the limelight. When a new governor is sworn in in January 2003, Schweiker said he'll head back to Bucks County. "Kathy and I are waiting for the peace and calm associated with our private life."

But Steve MacNett, general counsel for Pennsylvania State Republicans, noted that Schweiker, at 49, is still a young politician.

"This probably increases his ability to return to elected politics down the road with a more established public identity than there might have been beforehand," MacNett said.

President Bush was also drawn to the drama of the mine rescue. Eight days after the rescue, while in Pittsburgh for a fund-raiser for gubernatorial candidate Fisher, Bush met with all nine miners for 20 minutes at the Green Tree Volunteer Fire Department.

"What took place here in Pennsylvania really represents the best of our country, what I call the spirit of America, the great strength of our nation," Bush told reporters after the meeting.

"It was kind of crazy," Phillippi later said of the encounter. "He's the president, but he was worrying about us."

As for preserving the memory of the incident, the nearby Windber Coal Heritage Center plans to add a permanent exhibition on the Quecreek rescue, said Chris Barkley, the center's director.

The center, which features exhibits of a model mine community and how early miners lived, is attempting to obtain the equipment used to communicate with the trapped miners, as well as other objects involved in the rescue and news articles from newspapers around the country.

The Smithsonian's National Museum of American History in Washington, D.C., inquired about obtaining the rescue capsule for display, but was told by the National Mine Health and Safety Academy in Beckley, West Virginia, that until a copy of the device is made, the original must remain there in case it's needed for another rescue.

John Weir, the Black Wolf Coal Company spokesman, said he's working to get a new rescue capsule made that includes some modifications the mine academy wanted anyway. That might allow the original to go to the Smithsonian.

The Smithsonian is also planning to ask the miners for their hard hats and the bucket in which they placed notes to their loved ones.

Charles Fox, director of the Somerset Historical Center, just a half a mile from the rescue site, said he's been asked about what place the accident will have in history. But he declined to answer.

"We're all rushing to put historical significance on an event that's still happening," he said. "The same thing happened with September 11. But also like September 11, this event has reminded people of what's important: relationships, friends, family."

Fox said the miners set a good example when they were in their darkest moments: "They kept their wits about them, worked together, and decided to survive or fail as a group. And no one above ground was asking 'Who's going to pay for this?' or 'What am I going to get out of it?' They just did it because it needed to be done."

In getting the job done, however, the rescuers didn't just save nine men, and the miners didn't just save themselves.

Together, they delivered to the public the kind of happy ending it so rarely gets to savor.

JOHN BEALE/POST-GAZETTE

EPILOGUE

MEETING THE PRESIDENT

EXCERPTS OF PRESIDENT BUSH'S REMARKS ON MEETING THE MINERS AT THE GREEN TREE VOLUNTEER FIRE HOUSE, AUGUST 5, 2002

Thank you very much. Governor, thanks for your introduction. Thanks for your calm in the midst of crisis. I want to thank you and all the good folks here in the state of Pennsylvania who rallied to save the lives . . . of nine valuable citizens.

Today we're here to celebrate life, the value of life, and as importantly, the spirit of America. I asked to come by to meet our nine citizens and their families because I believe that what took place here in Pennsylvania really represents the best of our country, what I call the spirit of America, the great strength of our nation.

So I want to thank you all for coming, for giving me a chance to come and share with you the optimism and joy of an historic moment. . . .

And so to the first responders here, I want to thank you for your spirit. For those who volunteered hour after hour to save a fellow citizen—somebody you didn't even know, but were willing to sacrifice on behalf of that citizen—thank you from a grateful nation. Thanks for the example you set. Thanks for showing our fellow citizens that serving something greater than yourself is an important part of being an American.

I truly believe the effort put in will serve as an example for others in a time of crisis. The spirit of America, the best of America was represented by those who spent long hours worrying about the lives of their fellow citizens.

The best of America was also represented in the technology and know-how of our mine safety folks—those who, on a moment's notice, used their skill to devise a way to save life. . . .

The spirit of America can best be seen with the families who are here. . . .

I appreciate and I know your dads and your husbands appreciate much more than me the fact that you—the energy you spent on seeing that they came out of that hole alive was an important part of rallying the country.

And that's another part of the spirit of America I want to herald, and that is the prayers that were said by thousands of your citizens. . . . And thank God the prayers were answered.

But most of all, the spirit of America was represented by the courage of the nine—nine folks trapped below the earth. They had one sandwich and two bottles of water. These are people that found an unshakable faith, not only in their fellow citizens and their families who would be pulling for them, but faith in God. These are men who, as Harry Mayhugh put it, "I just didn't see how we were going to get out." That's what he said, "I don't see how we're going to get out." But he said, we're going to—we've got to pull together to get out.

In other words, they understood that they needed to rely upon each other, rely upon the strength of each. They huddled to keep warm, they said prayers to keep their spirits up. They understood they needed to tie together to fight the underground current. It was their determination to stick together and to comfort each other that really defines kind of a new spirit that's prevalent in our country, that when one of us suffer, all of us suffers; that in order to succeed, we've got to be united; that by working together, we can achieve big objectives and big goals. . . .

It's an honor to be here with you today. I want to thank you for the example you set. May God bless you, may God bless your families, and may God continue to bless America. Thank you very much.